Morris B. Parker's
WHITE OAKS

Life in a New Mexic

Tucson, Ariz

Morris B. Parker's
WHITE OAKS

Gold Camp, 1880-1900

Edited with an Introduction by
C. L. Sonnichsen

E UNIVERSITY OF ARIZONA PRESS

About the Author . . .

Morris B. Parker was not quite eleven when he moved with his family to White Oaks in 1882. He grew to manhood there, worked variously in the mine, the mill, and as an assayer, finally leaving when it became obvious he would have to make a career elsewhere. In the twenty-odd years covered in these reminiscences, Parker saw White Oaks develop from a crude settlement into a booming mining camp, then just as quickly decline as the mines petered out. Subsequently he pursued an adventurous life as a mining engineer, most of it in Mexico where his friends ranged from Pancho Villa to Porfirio Díaz.

About the Editor . . .

C. L. Sonnichsen, distinguished Southwestern historian, folklorist, and author, is a self-styled "grass-roots historian." The holder of the Ph.D. degree from Harvard University, he became a faculty member of the University of Texas at El Paso in 1931, serving as chairman of the English Department and graduate dean before becoming Harry Yandell Benedict Professor of English in 1966. Dr. Sonnichsen has received numerous honors for his scholarship and writing. His many books include *Cowboys and Cattle Kings; I'll Die Before I'll Run; Roy Bean: Law West of the Pecos; The Mescalero Apaches;* and *The Pass of the North,* for which he received a Texas Institute of Letters Award in 1970.

THE UNIVERSITY OF ARIZONA PRESS

I.S.B.N.-0-8165-0261-7
L.C. No. 75-143274

Acknowledgments

THE EDITOR of a historical work accumulates obligations as a librarian accumulates books. Without them he would not be in business. First thanks in the present case go to Morris Parker's daughter Lina Mathews, who provided the manuscript and much of the supplementary material. Her interest and help were invaluable. Likewise indispensable were the advice and counsel of John Kelt of Tucumcari, an old White Oaks boy who has made the history of the place his hobby for many years. He has accumulated a wonderful file of stories, photographs, and maps relating to the town, has kept up with most of the old inhabitants or their descendants, and has been most generous in sharing his wealth.

Of first importance also has been the assistance of Dr. Harwood P. Hinton, editor of *Arizona and the West* (who first suggested the need for a book on White Oaks); Paul W. Payton, editor of the *Lincoln County News* (Carrizozo); Xanthus Carson of Albuquerque, historian and writer; Mrs. Virginia Lee Starkey, reference librarian of the Historical Society of Colorado.

Carrizozo is a history-minded town. Even the high-school students, with the encouragement of Superintendent Glen Ellison and his staff of teachers, take pleasure and pride in keeping alive the old days in Lincoln County. In a community like that, everyone wants to help. Old residents like Mrs. Clara Snow, Mrs. Nettie Lemon, Herman Kelt, Jess Fulmer, Truman Spencer — all the rest — are ready to talk. Mr. and Mrs. C. R. Wetzel, who have brought hospitality to the Hoyle House, have memories and pictures. Mr. and Mrs. Pat Murphy, in the Gumm house across the valley, have been just as helpful. Mrs. Jackie Silvers has treasures in her House of Old Things at Ancho which are of great interest to historians.

Among the rewards of working on Mr. Parker's manuscript, not the least is the pleasure of knowing these people, their fellow townsmen, and their town.

Former residents of White Oaks, like Mrs. Lorena Sager of Santa Fe and Mrs. L. R. Cadenhead of El Paso, never get tired of talking about the old town and sharing their pictures and relics. A hearty thank-you to all of them.

C. L. S.

Contents

Liveliest Ghost Town in the West *1880 - 90+*

East side of state near Carrizozo

WHITE OAKS, NEW MEXICO, is always spoken of as a ghost town, but in some ways it is more alive than it was during the eighties and nineties when it was a booming gold camp with as many as 2,500 inhabitants and dreams of future greatness.

True, it has been left to decay on a dirt road that leads nowhere in particular twelve miles from Carrizozo, county seat of Lincoln County, and ten miles east of the turnoff from Highway 54. The tourists and summer colonists at Ruidoso and the racing fans at Hollywood on the far side of the mighty mass of Sierra Blanca have seldom heard of it.

True also, by 1970 not much was left of the old town in the way of brick and lumber and adobe. Only a few crumbling, windowless business blocks stood beside the once-busy main street. The venerable two-story school-house had been disused for decades. Only two of the big houses were left — the white-frame castle of the Gumm family (belonging to Mr. and Mrs. Pat Murphy) on the north side of the valley, and the Hoyle mansion, the home of Mr. and Mrs. C. L. Wetzel, overlooking the town from the south. The Hoyle house is such a fine example of faded Victorian elegance and such an unlikely thing to appear in the New Mexico countryside that people have felt obliged to make up stories about it.[1] The best one says that Watson Hoyle built it for his prospective bride, and when she wrote him that she was not coming, he walked up to the North Homestake and leaped to his death down the main shaft, over 1,000 feet deep. Actually Hoyle lived in the house for years, as Morris Parker tells the story, and left it only when he thought he could do better elsewhere.

Half a dozen families, notably the Bud Crenshaws, sparkplugs of the annual White Oaks rodeo, still inhabited less pretentious homes, but they sent their children by bus to the schools in Carrizozo and did most of their shopping there. There is no denying the fact that White Oaks was gone — destroyed, torn down, moved out.

Well, then what is this talk about its being more alive than ever? The answer is obvious. White Oaks has become a legend, larger than life, a golden memory in the minds of people who remember it in its heyday, in the minds of their children and grandchildren who have grown up listening to the old tales, in the minds of a large percentage of the citizens of southeastern New Mexico. They all cherish it as their own special and personal part of

---

[1] For variations on the theme of "Hoyle's Folly" see the *Lincoln County News* for May 26, 1969; Marjorie White, "Aura of Golden Years," El Paso *Times,* August 24, 1969; John Sinclair, "Little Town of Heart's Desire," *New Mexico* 18 (December, 1940): 18; Nell Murbarger, "Ghost of Baxter Mountain," *Desert Magazine* 15 (November, 1952): 4-8.

American history, the past still present, a portion of their heritage which they would not willingly let die.

The phenomenon is not unfamiliar. The West is full of towns that didn't make it — real ghost towns cherished by none, going back to the dust with nobody to care. But sometimes one of these fragments of the Old West refuses to die. Its history need not have been specially colorful or violent. All it has to have is a collection of former citizens who love it and keep it alive in their hearts. Such people are joined to each other by ties stronger than brotherhood (at least some brotherhoods). They keep up with each other; they write, or help to write, articles and books; they get together at the slightest excuse. Sometimes they go on pilgrimage back to the old town — or they have a picnic and reunion in a park at Long Beach, California, or some other place near where they live in exile. When they drop off, their sons and daughters carry on, while high-school students, local historians, and curious outsiders ask questions and take notes and help to keep the old days at Tombstone or San Marcial or Tascosa or Cabezon alive in the hearts of their countrymen. Mrs. Nettie Lemon said it for all of them, speaking of the men and women who were classmates in the White Oaks school in 1906: "Our love for White Oaks and each other is very dear to us."[2]

Nettie and her schoolmates were quite right in regarding White Oaks as a remarkable place. For one thing, its gold mines produced four and a half million dollars in twenty years or so of activity. It was on the edge of the battlefields of the Lincoln County War, and Billy the Kid was a familiar figure on its streets in 1880

---

[2] White Oaks cemetery dedication, May 30, 1969.

and 1881. Pat Garrett was the sheriff in the early eighties. Shady characters moved in and out, and there were murders and hijackings.

There is some disagreement about how tough White Oaks really was. F. Stanley thinks it was a headquarters for rustlers, a town where "a galaxy of gunslingers, horse thieves and gamblers" flourished; where "it was almost a ritual to hurrah the town every night."[3] A writer in *True West* declares that by 1895 White Oaks "boasted . . . as many bawdy houses with painted 'fancy' women as the rugged miners would support — and that wasn't a few. The largest casino was operated by a canny dealer named Madame Varnish, so-called, and justly entitled, because of her 'slick' way of dealing with the miners."[4]

There were violent episodes, of course, particularly in the early years, but White Oaks was no frontier Sodom, as Morris Parker makes quite clear in the present volume. The forces of good and evil never did battle it out in the saloons and on the streets of White Oaks. Gamblers and good-time girls never cut much of a figure there. It was, in truth, a pretty civilized place with churches and schools, literary societies and dramatic clubs, educated immigrants from the East, gentlemen and ladies who knew their way around in the world. It was not Cambridge, Massachusetts, of course; but neither was it Bridger's Wells or Poker Flat. The boys went off to college and came back to marry the local girls. It sounds almost too normal and healthy to be true.

Visitors were invariably astonished at the high level

---

[3] F. Stanley, *The White Oaks, New Mexico, Story,* published by the author, n.p., n.d.

[4] Xanthus Carson, "When a Lincoln County Ghost Town Roared with Life," *Lincoln County News,* May 28, 1969 (reprinted from *True West*).

of White Oaks social life. On the evening of July 9, 1892, Charles Longuemare, editor and publisher of the El Paso *Bullion* (which featured on-the-spot reporting of mining developments in the Southwest), attended a party given by J. Y. Hewitt for 200 guests. "Professor" Longuemare found his vocabulary almost inadequate to describe his feelings as he listened to selections from light opera and "the dear old arias from the old masters," and as he observed the elegant dresses of the ladies. "We do not exaggerate," he declared,

when we make the statement that the beautiful toilets of the ladies present would be difficult to equal in any Southwestern city, El Paso included. . . . I will close my remarks upon the local conditions of White Oaks by saying that during my stay I did not see a single revolver in sight, that peace and prosperity were visible everywhere, and that as usual the editor of the *Bullion* found a hearty welcome and a kind greeting from all he met.[5]

The noticeable percentage of immigrants from the North and East may have made the difference. Maurice G. Fulton in his account of the Lincoln County War notes that White Oaks was never "hospitable to Billy the Kid" and continues as follows:

The people of White Oaks mostly came from an Eastern environment that had made them sensible and law-abiding. Saloons, gambling halls, and other means of recreation and stimulation . . . they would tolerate; but they did not intend to let White Oaks become a rendezvous for . . . "murderers, horse and cattle thieves, and escaped convicts."[6]

Visitors, even in the earliest times, never failed to

[5] *The Bullion,* July 11, 1892. A writer for the El Paso *Times* reported on April 23, 1887, that "there is not a single bawdy house and only five saloons in a population of 500."

[6] Maurice Garland Fulton, *History of the Lincoln County War,* edited by Robert N. Mullin (Tucson: University of Arizona Press, 1968), p. 380.

take note of the cosmopolitan character of the White Oaks aristocracy. In 1885 a representative of the *Rocky Mountain News* came back to Denver with this account of the New Mexico gold camp:

> In twenty-five years of travel on the frontier and among the mining camps, I have never failed to meet people of intelligence and culture, who seem strangely out of place in the remote districts where circumstances have cast their lives. White Oaks is no exception. . . . Chosen to superintend some great mining enterprise, they have come to the frontier, brought their families with them, and made homes where fortune has located them. Their influence and the influence of their wives has had an ennobling effect on pioneer life and aided in the molding of frontier society into more refined, cultured and virtuous channels. These observations are suggested by remembrances of pleasant evenings in the parlors of Mr. E. W. Parker, the manager of the South Homestake Company. A gentleman of national acquaintance, familiar with the best phases of social and official life at Washington and all the great cities of the East, once in charge of all the great mail routes of the West, with a broad experience with men and affairs, fond of the society of friends and hospitable to the extreme, the visitor finds in his residence a courteous gentleman as a host, a model wife and mother as hostess, and in the informal but refined hospitality which is proferred to all guests notes the admirable influences which radiate from such a home.[7]

E. W. Parker and his wife were the parents of Morris Parker, the author of this book.

Like Saul of Tarsus, a resident of White Oaks was "a citizen of no mean city," but White Oaks had one distinction that Tarsus could not match. It committed suicide.

Of course many factors were involved. When the

---

[7] From Mrs. Morris Parker's scrapbook, exact date not given. The scrapbook was, at the time of writing, in the possession of Morris Parker's daughter, Lina Parker Mathews.

mines played out, the town, as we used to say, "went down." But it would unquestionably be a flourishing little metropolis today if the leading men just before the turn of the century had not gone out of their way to cut their own civic throats. They did it because of the railroad.

For years they had waited for the rails to arrive and give them permanent status. The original surveys called for the line to be built through White Oaks, and the inhabitants became convinced that they had not merely the best route but the only route. When the railroad impresarios asked for land for right-of-way, railway station, division offices and shops, plus a cash subsidy, the village elders said no. They would not give anything to a corporate octopus which would have to come to them anyway. White Oaks gossip adds that a prominent cattleman whose acres the rails would have to cross decided that the locomotives would not do anything for him and probably would scare his cows, so he refused to negotiate with the railroad officials. Another story says that no less a figure than John Y. Hewitt, the most important man in White Oaks, was one who put such a high figure on land inside the community that the negotiators simply walked off and told the surveyors to forget White Oaks and head for Corona.[8]

So White Oaks was left high and dry. A new town, Carrizozo, was established out on the prairie west of Carrizo Mountain. It became the county seat of Lincoln County and the place where everybody went to see the doctor or buy groceries or sue somebody. Children began to travel from outlying districts (including White Oaks)

[8] Charles L. Wetzel, "Mining in White Oaks, New Mexico," Ms.; Mrs. C. L. Wetzel to C.L.S., January 4, 1969.

to go to the Carrizozo schools. Soon White Oakers began moving to the new town, where opportunities were better, and some of them brought their buildings with them.

It is said that for a while the "new" people in Carrizozo and the "old" citizens from White Oaks tended to view each other with some lack of cordiality. If this was once true, by the late 1960's it was true no longer. White Oaks had become the showplace of the region, right on the heels of the old town of Lincoln; and all good men and true (including the entire male population of Lincoln County) worked to make the most of it and keep it from going all the way back to the earth from whence it came.

It took the White Oaks cemetery project to show what a concerted effort the whole county was capable of making. The top officials of New Mexico agreed to make a state monument out of the venerable burying ground and the dedication was set for May 30, 1969. With Mrs. Clara Snow of Carrizozo as chairman, a committee was formed in 1968 to plan the celebration.

Mrs. Snow had no trouble getting all the help she needed. Since 1969 was the centennial year for Lincoln County, many wheels were turning and people were already at work. Interested citizens in Alamogordo, Tularosa, Lincoln, and other neighboring towns took part in the planning and preparation, and before long Mrs. Snow had a program set up. Truman Spencer, colorful representative of one of the old ranching families of the region, agreed to serve as master of ceremonies. Lieutenant Governor E. Lee Francis promised to make the principal speech. Arnold Boyce, band director, said he would produce the Carrizozo High School band for some special music. And word went out to former residents, scattered all over the country.

The result was the biggest reunion of old settlers in White Oaks' history. The old schoolhouse was furbished up for the pot-luck luncheon, and a big crowd was there at noon on May 30. It was a noisy and enjoyable affair with everybody greeting old friends and reminiscing with everybody else. If all of it could have been recorded, a comprehensive history of White Oaks would have been the result.

"Yes, Bill Gallacher's father died when he fell down the shaft of the North Homestake." (Bill Gallacher became the patriarch of an important ranching family.)

"There were four rooms in this schoolhouse — two upstairs and two down. This one was made into a playroom. We used to roller skate in here."

"One thing about this school, you had to make a seventy-five average or better to get out of the grades. You could make two grades a year if you were smart enough."

"We had a baseball team here, a bunch of us kids, fourteen or fifteen years old. We played the miners here in town and we made 'em hard to catch."

John Kelt, a native son who in 1970 was getting ready to write a history of the town, brought a picture of Dave Jackson, the kindly Negro who lived on in White Oaks long after it had fallen into decay and who became a sort of focus for White Oaks memories.[9] His name came up often in the clatter of talk.

"Dave Jackson?" (Frank Williams is talking.) "He

[9] For years Dave was the focus of an annual Dave Jackson Day with dinner at the old schoolhouse ("Ramblin' around Lincoln County," tape-recorded interview with Dave Jackson, the last of three, in *Lincoln County News,* April 27, May 4, 11, 1956).

named every one of us — give us a nickname. Any time there was any sickness or anything, that was the first guy we went to. I'll never forget one night. We lived in this old Taylor house, right over here. It was along about two o'clock in the morning, my daddy got real sick. I don't know what it was — pneumonia or something. I can remember I went over and tapped on that glass and I said, 'Jack, my daddy's sick. Come quick.' Boy, he was there, and he stayed right there too."

After lunch the crowd moved to the cemetery, a mile back on the Carrizozo road, where old White Oaks lies sleeping peacefully on the hillside, the history of the town since 1880 displayed on the tombstones. W. C. McDonald is there with his family, the first governor of New Mexico after statehood; John Wilson, who helped to make the gold discovery; Ah Nu, the beloved Chinese restaurant keeper and laundryman; Mrs. Susan McSween Barber, survivor

WHITE OAKS CEMETERY DEDICATION
Lt. Governor J. Lee Francis and Truman Spencer

— *Lincoln County News*

CATTLE QUEEN
OF NEW MEXICO

*Susan McSween Barber
of Lincoln, Three Rivers,
and White Oaks*

— Lina Parker Mathews

of the Lincoln County War who became famous as "The Cattle Queen of New Mexico"; the miners who lost their lives in the Old Abe fire in 1893 — all in a row with cedar posts at their heads; Dave Jackson himself, with his wife Mary, resident at last in the cemetery he cared for conscientiously until age and disability intervened.

Since Dave was no longer there to do the job, the graves and grounds had been refurbished by a small army of workers. Glen Ellison's students from the Carrizozo schools made repeated visits with hoes and rakes, and others pitched in to help. On the big day the place was clean and many of the graves were decorated. Perhaps 200 people were assembled when Truman Spencer introduced his platform guests and the program began. Loyal Craig told how the Knights of Pythias had established the cemetery in 1880. Lieutenant Governor Francis expressed his good wishes. And then came what many considered the high point of the celebration — reminiscent talks by four old residents of White Oaks — Nettie Lemon, John Kelt, Clara Snow, and Roy Harmon.

"In the early days Memorial Day was an all-day observance," Nettie Lemon told the crowd. "It was run by the G.A.R. They came from all over the county for their reunion. They would meet at the Congregational church, form their lines, and led by our town banker, they would march to the cemetery. They would hold their services, place the flag on each veteran's grave, sound taps, and then go back to town to a bountiful dinner the ladies had prepared. Then after dinner they would sit around on the sidewalk smoking their pipes and fight the Civil War all over again. And sometimes it got a little bit rank."

The speeches over, the crowd drifted over to the Hoyle house, opened for the occasion by the present occupants, Mr. and Mrs. C. L. Wetzel, and there were refreshments and more reunions.

As the day departed, so did the guests, having proved once more that White Oaks was still alive and flourishing.

This was one reunion that Morris Parker had to miss, though he would have enjoyed it. He died at his home in Hermosa Beach, California, on October 18, 1957, after a long and adventurous life as a mining engineer. He left White Oaks when it became obvious that he would have to make a career elsewhere. Until about 1935, he lived and traveled about in Mexico, examining mining properties, old and new, in many an out-of-the-way place. He took his family with him. Mrs. Parker was often the only woman in camp. Three of his five children were born south of the border. He outwitted bandits who wanted to capture him and hold him for ransom. Once he was a captive of the Yaqui Indians. Another time he contracted the terrible mountain fever and made it on muleback across the mountains 150 miles to civilization and medical help. Pancho Villa managed mule trains for him. Jim Douglas, for whom Douglas, Arizona, was named, was his associate. The great men of Mexico — Madero, Orozco, Carranza, Huerta, Porfirio Díaz, Luis Terrazas — were his friends.

Reluctantly, like many another American who had invested much of his life in that country, he left Mexico for good when the Revolution made longer stay impossible. For a while he lived in El Paso, Texas, for the sake of his children's schooling, but his profession took him to Arizona, to Oregon, to California, and finally to Alaska. When he retired, he settled at Hermosa Beach, California, where he and his family had spent their summers for a good many years. Mrs. Parker died in 1955, and he followed her two years later.

He was a literate and thoughtful man, and he left a good deal of writing behind him, including the manuscript of this book, but he was too modest to think seriously of publication. He did hope that his stories might interest some of his long-time friends still living in the White Oaks neighborhood, and he sent a copy of his memoirs to circulate among them. I read it when it came into the hands of C. E. Burns, druggist and historian, of Alamogordo (later of Carrizozo), who put me in touch with the author.

"For your information," Mr. Parker wrote me on March 31, 1948, "would say the manuscript is composed largely of reminiscences — from memory, diaries and old notebooks — written primarily with the thought that maybe 50 or 100 years from date one of the children or grandchildren may be interested; wonder what granddaddy did when he was a kid."

It was obvious, even then, that somebody besides the grandchildren might be interested in these memories, but it took another twenty years for anything to be done about it. Harwood Hinton, editor of *Arizona and the West*, pushed the button. In the course of a conversation in his office one morning in the summer of 1968, he remarked that one of the gaps in historical writing about the Southwest was the story of White Oaks. "I wish somebody would do a book about it," he said.

"Morris Parker did a book about it," I told him, and we went on from there.

The manuscript does more than fill a gap in the chronicles of the mining frontier. It opens a door on a special time in our history which we look back on with fascination and pride — the day of the first comers. As we lose more and more of the attributes of this earlier time, we recall it and relive it with nostalgic pleasure, and we are grateful for anything which places it in perspective and brings back its rich overtones and special flavor.

Morris Parker is the ideal chronicler of this vanished world. He had keen observation, tremendous recall, a wry sense of humor, and a love for his fellow man which he retained even when he was recounting his fellow man's peccancies and peculiarities. He set it all down with a relaxed charm which suits his material and beguiles his reader.

He did not think it necessary to fictionize or invent.

White Oaks did not need to be improved. He told about it just as he remembered it — conversationally and informally. And from now on anybody who wants to know what life was really like in a Western mining town can pick up his book and find out.

C. L. SONNICHSEN
El Paso, Texas

White Oaks died because the people refused to let a RR come through. Carrizozo was built & accomodate the RR & many people went to live there.

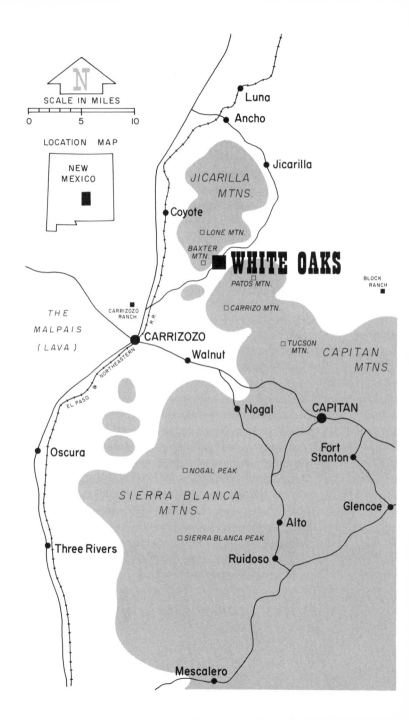

From New York Grapes
To White Oaks Gold

MY FIRST SIGHT of New Mexico occurred when I was nearly eleven years old in June, 1882.

The Santa Fe railway was then building west from Atchison and Topeka, Kansas, toward Los Angeles, but it had not yet reached Albuquerque. The passenger terminal was Las Vegas, New Mexico.

From Las Vegas our family drove overland by team and wagon 175 miles almost due south to White Oaks, where gold had recently been discovered. We expected to make a hundred million or so by getting there first and being on the ground floor. We did not quite realize our expectations in yellow metal, but we did find a wonderful playground for growing boys. We did not know that we were leaving our old home for good and we did not consider ourselves as pioneers. In later years, however, we realized that we were first comers during a most interesting period of development in the great Southwest.

White Oaks, our new home, was for two decades prior to 1900 a busy, bustling mining town where a lot of West-

ern history was made. It is a ghost town now. The charm
of its natural surroundings is unchanged, but the busy
streets are deserted and the old families have moved
away or gone to rest in the grass-grown cemetery. All
that is left is the memory of a great experience in the
minds of us who lived there before the turn of the century.

Father (Erastus Wells Parker — known as E. W.)
was a Westerner — in heart, spirit, and work. He was the
only son of a well-to-do lumber merchant of St. Louis, but
his own great interest was horses — fast horses — and he
is said to have shown some high steppers on the boule-
vards of his boyhood hometown.

The family came from Dansville, New York, where
Father was born, in 1852. His father set up a sawmill on
the Mississippi, floated logs down the river, accumulated
a fortune before and during the Civil War, and lost the
greater part of it during the panic of 1873.

Mother (Emmeline Brown) lived with her folks at
Penn Yan thirty-five miles east of Dansville at the north
end of Lake Keuka, one of the "Finger Lakes" in the western part of New York state. Her father, Morris Brown,
after whom I was named, was a prominent land owner,
attorney, politician, and judge. The Browns were among
the aristocrats. Mother was a "society belle," and also a
talented pianist and singer.

Father and Mother were married at Penn Yan in
1866. For a wedding present, his father gave them a beautiful, well-cultivated vineyard on the west shore of Lake
Keuka about five miles south of Penn Yan. The place was
later known as Parker's Landing. Farming and marketing
Catawba grapes, however, was not Father's idea of right
living. He was not accustomed to the work; he missed his
former associates; he found country life slow and monot-

onous. By making frequent trips back to St. Louis he retained contact with his Western friends and led a sort of double life for seven years, cultivating his grapes in New York state and spending as much time as he could in his Western home, where life was far more attractive.

During those seven years four children were added to the family circle: three boys and one girl. The little sister passed away a few weeks after birth. The boys, James, Frank, and Morris, were five, three, and two years of age when they left the farm in 1873.

The change came when some of Father's friends in Atchison, Kansas, top officials of the National Mail Company, offered to make him manager of the Southern Division. He accepted pronto.

The National Mail Company was one of the large stagecoach outfits carrying mail for the U. S. government along with express, baggage, and passengers, covering the entire country before the advent of the railroads and during their period of extension.

During the seventies Father was away from home most of the time, traveling, supervising, checking routes then in existence and establishing new ones, particularly west of the Mississippi River. Apparently he knew all the roads, many of the trails, and a lot of territory where neither trails nor roads existed. He was one of the best-known and most widely traveled Western men of that period. Few men had as great an opportunity to observe the rapid growth of the mining industry and the quick wealth resulting from the discovery of gold and silver outcrops in the West. The idea of trying his luck in the mining game developed in his mind. "It looks easy — why not me too?" Then White Oaks happened.

There was some history behind it. Gold in Lincoln

County was first discovered in the Jicarilla district ten miles northeast of White Oaks. During the 1860's Miguel Otero, father of the Territorial governor of that name, built an earth dam near the head of Ancho Gulch to impound flood waters from summer rains for use in sluicing placer gravel below the dam.[1] Operations are said to have been successful, one season's work producing $60,000. No commercial deposits or veins in solid rock have been found in that district.

The first discovery "in place" (in solid rock) where exploitation was profitable was the American mine near Nogal, sixteen miles by road southeast of White Oaks. It was made in 1866 by Billy Gills inside the Jicarilla Indian reservation, which at that time adjoined the Mescalero reservation to the south and extended north to include the Jicarilla and White Oaks districts. Working from a hideout cabin, Gills operated a single *arrastra* so effectively that, with the help of some rich specimens, he realized sufficient funds for the purchase and installation of a light, 750-pound stamp mill which intermittently for thirty-odd years provided an uncertain revenue. Pockets of high-grade ore and promising specimens were always an alluring incentive, but during its many years of operation, the property was never developed or explored beyond a depth of 250-300 feet.

It was because of the two discoveries mentioned that Congress was persuaded to abolish the Jicarilla reservation and move the Indians elsewhere, thus throwing open the region to settlers and prospectors.

---

[1] Socorro *Bullion,* July 1, 1884: "It is well known that the Rev. B. Bernard, the late pastor of Socorro, and Don Miguel A. Otero some twenty years ago worked a placer there successfully."

Gold discovered in 1870's

About a mile west of present-day White Oaks, gold was discovered by a group of prospectors — presumably from the Jicarillas — in the placer gravel of Baxter Gulch during the late seventies.[2] The vein outcroppings were uncovered shortly thereafter, in 1879.

Four names are of record: Charles Baxter, after whom the gulch and the mountain are named; John E. Wilson and his partner Jack Winters; and George Wilson (no relation to John), the outcrop discoverer, an outlaw on the run from Texas authorities.[3] George sold his find to the two partners, Wilson and Winters, for a couple of silver dollars, about two ounces in gold dust ($38.00), and a small pistol; then left for parts unknown.

The partners located a standard mining claim, the Homestake, 1,500 feet long by 600 feet wide — twenty acres. They then severed partnership and split the claim, Wilson taking the south half, Winters the north. Wilson then added 750 feet to his holdings, thus covering 1,500 feet south of the partition line.[4] Winters was satisfied to retain his half claim, 750 feet, adding nothing. The two properties thenceforth were known as the South Homestake and North Homestake, respectively. The customary new-gold-discovery stampede followed. Within a short

---

[2] Mexican residents of the district had known about the placers for some time. Fayette A. Jones, *New Mexico Mines and Minerals* (Santa Fe: The New Mexican, 1904), p. 172; Stanley, *The White Oaks Story,* p. 1).

[3] Another version of the Wilson story says that he was an ex-member of Quantrill's raiders and had to keep ahead of Yankee pursuers (Wetzel, "Mining in White Oaks").

[4] For fuller accounts of these proceedings, see Stanley, *The White Oaks Story,* quoting the Albuquerque *Journal* for April 6, 1881; the El Paso *Times,* April 13, 1887; C. E. Stivers, *White Oaks, New Mexico* (El Paso: Press of the Daily News, 1900).

time all of Baxter Mountain and a goodly portion of Lone Mountain were plastered with locations.

At the time of the discovery, White Oaks was a long way from anywhere. The main line of travel south from Las Vegas and Santa Fe to El Paso (thence into Mexico) stuck to the banks of the Rio Grande. The U.S. Mail stage route did the same, following the old Spanish trail. San Antonio, on the west bank of the river ninety miles west of White Oaks, was the nearest post office. After the discovery of gold in 1879, a post office was applied for. The petition was granted and the mail contract was awarded to the National Mail Company.

Ostensibly on business pertaining to this company, Father drove from San Antonio to White Oaks. One look at the rich gold specimens, and the virus took effect. With no knowledge of practical mining and no conception of the unreliability of high-grade ore specimens, Father picked up a handful of ore which was two-thirds pure gold and worth $400,000 per ton. He lost no time in becoming the owner of the South Homestake.

Had Father known anything about mining, his vision might, in a measure, have become reality. His three boys, brought up in practical mine work, educated in the profession, have spent a lifetime hoping and searching for a place with somewhat similar conditions. Thousands of others have had the same lifetime objective. Father, of course, had no idea of the difficulties involved. It seemed to him that capital was the key to everything. Speeding to St. Louis and resigning his connection with the stage and mail company on the way, he interested his uncle Erastus Wells, after whom he was named, and the two of them took immediate steps to flood the U.S. Mint with gold bullion.

Starting from scratch in a region far removed from sources of supply, Father found that the undertaking was not as simple as it seemed. Everything, from frying pan to final retort and mould, was an integral part of the problem — to be planned and coordinated in advance as part of a completed workable structure operated by a personnel which knew what it was all about. A mistake could easily prove disastrous.

Ordinary working tools and supplies were available from Las Vegas but the more essential heavy equipment had to come from St. Louis, Denver, or Chicago. A preliminary list of such equipment, necessary to follow up the hand work and make a start, included the following:

Two portable sawmills with accessories
Thirty to forty span of oxen, with yokes and bows
Heavy logging wagons with chains and other equipment
A well-drilling outfit with cable and standard rigging
A pole-pipe boring outfit to make a wooden pipeline — for water
Mine hoist, cable, sheave, buckets, cars, rails, etc., with boiler, engine, iron pipe, fittings, blacksmith shop, and tools

Once these items were on the road, machinery, equipment, and supplies for a complete mill were ordered, to be shipped as soon as they could be assembled. The mill, of course, was the main object — and by far the biggest headache.

Most of this equipment was purchased in St. Louis, shipped by rail to western Kansas as far as the railroad extended, transferred to wagons, and hauled by mule and ox teams to Las Vegas, to San Antonio, to White Oaks. Time was a big item.

The sawmills were placed about ten miles east of

town near the north base of Patos Mountain in the center of a magnificent stand of virgin pine. Excavations and foundations for the gold mill were prepared on the hillside west of and below town. The mill water supply was brought to this point from White Oaks springs, two and a half miles east, by means of a three-inch wooden pipeline made of small pine trees bored, reamed, beveled, and joined together. The line was placed in a ditch and buried two feet below the surface.

Meanwhile work went forward on the necessary buildings — shops, barns, storehouses, and the like.

Two and one-half years elapsed before the mill was completed and an attempt was made to start production. It was a long, hard grind, requiring a lot of patience and money — especially the latter — but at last the end was in sight. The strain of anticipation was over and Father was confident of the result. What a celebration it would be to have the family present for the start! And why not? School vacation was about due. No better opportunity for the boys and their mother to rough it might ever occur. It would make the start perfect.

Father's suggestion was met by the unanimous approval of the four in St. Louis, and soon all plans were centered on the idea: Out West for the Summer.

Tenderfeet Go West

IN JUNE, 1882, the family was still living in St. Louis, but our time there was nearly up. For the past eight summers we had gone east to Penn Yan, Mother's hometown, but this year we were going West to spend our vacation.

We thought we knew something of what the West was all about. A couple of years previously, Father had sent us boys a small Indian pony from the Indian Territory (now Oklahoma) with all the paraphernalia: saddle, saddle blanket, bridle, and quirt. In addition he sent a pair of buckskin breeches, a jacket to match, moccasins, and a headgear complete with feathers. The pony was a beautiful animal with long, bushy mane and tail, full of pep but gentle as a kitten, and how he could run! It was great sport for one of us, in full Indian regalia, to gallop down an alley near our home yelling variations of what we thought was an Indian war whoop, and scattering a group of unsuspecting children. We were acting out the usual fireside tales of that period.

We had no idea of the difference between the real

9

West and the imaginary West of those fireside tales, but we were prepared for something bigger than chasing the neighborhood children. The thought of a trip into the unknown, no matter how inaccurate our conception of it, was the most exciting thing that had ever happened to us, and a halo of adventure rested upon the little train as we climbed aboard.

Kansas stands uppermost on my list of memories, still vivid. Our train was one of the first to cross the state. As we looked from the car window across the vast, flat prairie, there was not a plowed field or a fence or a farmhouse west of Newton, where we arrived after covering the first third of the east-west distance, or if there was one, it was the only one. Twenty-five years later, Newton was near the center of one of the most productive agricultural regions on the continent, famous for corn and wheat, the "bread basket" of the nation.

I watched the extension of the cultivated fields westward as I traveled back and forth between school and home during the next few years. The advance westward was like a flowing tide. First isolated windmills dotted the desert; then came farmhouses, fences, and cultivated fields. The population of Kansas increased five-fold between 1880 and 1905 — from 300,000 to 1,500,000.

Newton, when we first saw it in 1882, consisted of a combination station and restaurant (Harvey House), a dozen or so rough frame buildings, and several tents. Between Newton and Raton Pass on the border between Colorado and New Mexico stretched the Great American Desert, recent feeding ground of many thousands of buffalo — a monotonous, treeless, awesome yet fascinating prairie, mile after mile of it. It was all grazing land carpeted with buffalo grass and tall prairie grass, dotted with sage and other brush, punctuated by an occasional windmill and supporting a few scattered herds of second- or third-grade longhorn cattle plus a few bands of sheep.

The only town of significance in the 500-mile stretch between Newton and Las Vegas was Dodge City, cattle town and Indian trading post. In every direction it was hundreds of miles to the nearest settlement. Half a dozen buildings and tents surrounded the station on the north side of the river. The old town, and still the main one, was a couple of miles away on the south side.

Across this desert prairie ran a single-track clickety-clack railway, standard at the time — a far cry from the smooth-running, air-conditioned Santa Fe Chief of later years. Paralleling the roadbed on one side were telegraph poles, each with but a single arm which supported possibly half a dozen wires. The train was made up of cars attached to one another by a single oblong cast-iron link and pin, though the company had added the extra precaution of heavy chains. The leeway afforded by this connection, especially when starting and stopping, was sufficient to jerk a dozing or daytime-sleeping passenger from his seat, and the sudden jolt was hard to get used to. The platforms between the coaches were dangerous traps. To step from one coach to another across the exposed coupling while the train was in motion was to invite disaster. Safety for passengers and crew depended on keeping an eye on one's foot and a hand on one's hat.

Somewhere along this stretch is said to be the longest railroad tangent (level, straight track) in the United States. We diverted ourselves by standing on the rear platform and watching the two rails converge in the distance. When the train made one of its infrequent stops, we looked on as the brakemen changed the hand-controlled semaphores. The roadbed was still wobbly — soft, uneven, and unsettled. The desert soil, tamped between the ties, was the only ballast, and men were still at work on the right of way. Whenever the train passed a section gang at work on the roadbed, the workmen stood alongside and yelled like madmen, waving us on our way. The conductor told us they were Indians. Indignant and anxious to prove to him that we were neither inexperienced travelers nor tenderfeet, we asked him, "Where are their feathers?" He got out of that one by explaining that they were *tame* Indians.

At each station, often an old boxcar on a disconnected side track, was a big, red, elevated water tank and, close by, a windmill — the sole source of water for the railroad. Dense clouds of dust picked up by the moving train drifted between us and these dreary little settlements, mingling with the hot, flying cinders from the soft-coal–burning engine.

To the west, far in the distance, we saw our first real mountains: snow-capped Pike's Peak and the Spanish Twins, great masses lifted high above the floor of the prairie. As we approached the Sangre de Cristo range, the conductor told us we would go over the mountain ahead, up one side and down the other. When a second locomotive was attached, head on, to the back of the coach in which we were traveling, we decided he must be serious and prepared for a definite thrill. Nothing unusual happened, however, and I vividly recall the disappointment. True, from the car window we could see the valley and sometimes the tracks far below, but there was no noticeable incline. As the tracks wound mysteriously among the hills, the valleys suggested a long roller-coaster excursion — a short push up and a long slide down. The noise of the engine behind and the smoke and cinders from the one ahead received far more attention than the landscape. We saw nothing at all to get excited about. At the summit the push engine was unhooked and left behind, along with our hopes for a thrill. With marked disgust we informed the conductor that his train went downhill not much faster than it came up.

Mother was the first to notice the changed atmosphere, the clear, fresh, invigorating quality of the air. We quickly forgot our disappointment and began to think with anticipation of the things beyond the encircling hills.

Three and a half centuries after the arrival of that adventurous cavalier, Don Francisco Vázquez de Coronado, we descended into his valley. Our purpose was the same as his: the search for gold. The country had been ceded by Mexico to the United States only thirty-four years before.

Las Vegas looked like a metropolis, with perhaps 500 dwellings made of adobe and built close together. A typical Mexican village, its streets were unpaved and deep in dust. There were a couple of two-story adobe hotels. A saloon occupied every corner. Rooming houses, stores, and more saloons filled up the rest of the business blocks. Well-stocked mercantile establishments, wholesale and retail, were centers of activity, distributing supplies to ranches, mining camps, and settlements in all directions. Materials for railroad construction were very much in evidence: freight cars loaded with rails, ties, and other materials; heavy wagons, dump carts, plows, scrapers, tools; work animals, both mules and horses — all going southwest farther and farther as construction progressed in that direction.

We boys were too young to notice, but Las Vegas at that time was rated as one of the toughest, most lawless spots in the West. Tombstone, later nicknamed Helldorado, was the only place that could have claimed precedence as a rough town, and it was not very far ahead.

In a few hours it was on to White Oaks, 175 miles south. Mother and Father rode in a light buggy; we three boys (James, Frank, and myself) traveled with Uncle Marshall, Grandfather Parker's brother, in a light spring wagon. Two horses pulled each vehicle. Tied to the sides of the wagon, resting on makeshift platforms, were two twenty-gallon half barrels which held our water supply. Along desert roads such as ours, water was scarce and often a source of income, the customary charge being twenty-five cents for each animal and a dollar a barrel (forty gallons) for what was carried away.

The road was ill defined, only a trail in places connecting the water holes en route. At night we camped in the open, slept on the ground. Our route led us past Salt Wells, a dry lake bed from which the Mexicans gathered salt by scraping the incrustations into piles and shoveling them into wagons or sacks. There were but two habitations along our line of travel. The first was a solitary, one-room rock house, night camp for a sheepherder on the Spence-Maxwell ranch. We passed the herder — a Mexican — on the road. After a bit of conversation which neither side understood, he accepted the proferred toll, two dollars, and with a "Gracias, adios," turned to follow his sheep. We continued on the dim trail leading south.

We noticed that Father took precautions, choosing our camping spots carefully and not allowing smoke or any large fires. We boys did not know it then, but 1882 was a bad year for Indian outbreaks in both New Mexico and Arizona. A number of bands under desperate and dangerous chiefs broke loose from their reservations — notably Geronimo and Loco (Crazy). These were the worst in Arizona, but they were rivaled by others from the Mescalero bands just south of White Oaks who

attempted to join the Apaches in Arizona. More than one
hundred white settlers were killed in that year.

Our only firearms were a .45 caliber Colt's six-shooter,
which Father always carried, and a .44 caliber Winchester
rifle on the seat of our wagon. All shooting, however, was
taboo. As it turned out, we did not meet any Indians.

We traveled as fast as the horses could stand to go.
Since we could carry only a limited amount of grain in
the wagon, they had to "rustle" for grass at night, three
of them hobbled, the other tied fast at the end of a fifty-
foot rope not too far from camp. They were about played
out when we arrived at Jerry Hockradle's ranch on the
afternoon of the fifth day. Jerry and his extra-fat, good-
natured wife had a well in the Jicarilla foothills ten miles
north of White Oaks. They treated us well, but they did
not refuse the customary fee for water. A few hours later
we had our first glimpse of White Oaks.

A small stream some two and a half miles east,
bordered by a heavy growth of white-oak trees, gave the
place its name. At the time of our arrival it had a population of about five hundred, mostly men. Our house was a
two-room log cabin with flat mud roof. A frame lean-to
kitchen was attached at one end, making the house
L-shaped. Floors were of rough one-inch boards which had
dried in place, leaving wide cracks between. Pieces of tin
cans nailed to the floor covered the holes where knots
had fallen out. Our initial domicile was no worse and
not much, if any, better than the average.

In construction the house was like all log cabins. It
was made of small pine trees, peeled and cut to the right
length, notched at both ends and laid horizontally. A post
in the center of each room supported the roof. Between
the logs were splints of wood held in place by square-cut

nails. When the logs were in place, openings for door and
window frames were sawed out. Finally mud mixed with
straw was forced into all the cracks and left to dry. The
roof consisted of three or more inches of mud which, when
dry, was covered with soil well tamped by bare feet and
heavy wooden blocks.

Frequent repairs were necessary but not expensive.
The roof was the most vulnerable part. A hard or long
rain could play havoc with it, and at such times washtubs,
dishpans, buckets, stew pans, anything that would hold
water, were placed under the drips which at times became
small — and muddy — rivers.

In due time we learned that houses of this type were
prime refuges, with plenty of breeding space, for centi-
pedes, scorpions, bedbugs, fleas, and other pests more or
less common in the Southwest.

To Mother, gently nurtured amid other surroundings, the experience must have been awful, yet no one
ever heard her complain. I have often heard her say,
"Our Heavenly Father always has taken care of us, and
I guess He always will." It was her motto. For us boys,
of course, living so close to nature was fun.

For three years the log cabin was home, after which
we moved to a more commodious eight-room story-and-a-
half adobe house at the other end of town in the center of
a forty-acre pasture.

We found that we had just missed the biggest event
of its kind in all White Oaks history, and the gossips
around town were still talking about it. A few days before
our arrival, a colored man rode into town mounted on a
horse carrying the brand of a ranch not far distant.
Unable to give a satisfactory explanation of his presence
or to prove ownership of the horse, he was arrested,

— Lina Parker Mathews

FRONTIER OASIS — *The Parker Home in White Oaks*

charged with illegal possession of the animal, convicted forthwith, and sentenced by unanimous vote of the townspeople.

Since the town had no jail in which the prisoner could be confined overnight, he was lowered into a thirty-foot (dry) well then being dug near the center of town and just off the main street. A guard was stationed on the surface to prevent escape. In the morning the violator of Western principles, who had committed the unpardonable crime of being caught, was hoisted from his prison cell, hustled across town, and halted on the other side of the arroyo southeast of the village where stood a tall, leaning piñon tree with a large branch about twelve feet from the ground. There, in full view of the crowd, he was hanged. The tree was just inside the upper or west gate of the forty-acre pasture we acquired in later years. The mark of the rope, where the bark was rubbed down to the solid wood, was visible for many years — a witness to the grim realities of early-day law and order.

Mountain Paradise

White Oaks surrounded by mineral wealth

WHITE OAKS is situated in a wide, gently sloping valley surrounded by mountains. Trees enhance the natural beauty and help to provide an environment of extraordinary charm. Elevation above sea level: 6,500 feet. To the north rises Lone Mountain with deposits of iron, lime, gypsum, native sulphur, building stone. To the east, Mt. Patos, heavily timbered with pine forests and bearing deposits of bituminous and semi-anthracite coal. To the south, majestic Carrizo Mountain, 10,500 feet high. To the west, Baxter Mountain, least in elevation but greatest in importance since it contains the gold-bearing formations.

On the hills immediately surrounding the townsite are generous mixed growths of juniper, cedar, piñon, and scrub oak. The higher elevations, mountains and plateaus, are well covered with so-called Western (white) pine, suitable for timber and lumber, especially if it grows at elevations above 6,500 feet. Along the watercourses and in the valleys are found cottonwood, sycamore, white oak,

THIS WAY OUT
Main road from White Oaks, going West

— *Lina Parker Mathews*

ROUGH COUNTRY
The road across the malpais west of White Oaks

and (occasionally) walnut trees. The juniper, cotton-wood, sycamore, and cedar support large bunches of mistletoe. Only the cottonwood and sycamore are decidu-ous, shedding their leaves each year. The leaves of the scrub oaks change color, green and brown, twice in the year: green after the seasonal rains; brown during the dry spells between.

Smaller growth includes cacti of various types, some offering beautiful flowers and edible fruit, the latter highly prized by the Indians who know how to handle the sharp-pointed stickers; mesquite, whose pods provide fodder for deer and cattle; algeroba bushes producing thick clusters of red berries, the main ingredient in a delicious jam; sage, catclaw, and so forth.

There are two rainy seasons: the main summer season (late June till mid-September); another during December and January. The winter rains are usually mild and last longer than the summer downpours. Cloudy skies and gentle showers last twenty-four hours or more. The summer rains are hard, fast, and *individual*. During the winter there is snow on the mountains — sometimes a foot or two in the valley — lasting only a few days.

The hot period occurs during the latter part of May and on into June until the summer rains begin. San Juan's Day, observed in Mexico as the Rain Saint's day, often is surprisingly accurate as a register of the first rain.

During this season the mornings are clear and bright — cloudless until about ten o'clock. A small, gray, fleecy cloud then appears in the distance, gently floating in the deep azure of the sky. It grows larger and larger. Soon other similar clouds can be noted, all separated by stretches of clear sky, each growing independently in its own field of accumulating moisture. By noon each has taken to itself the form of a heaped, rounded mass, the horizontal base getting darker and darker, the sides and top a bright, glistening rim of gold. Several such clouds may be seen at the same time. From one or all of them rain may fall — direct, vertical if the air be still; slanting when the cloud is moved by a breeze — covering perhaps only a few acres, possibly a square mile or more. The outer edge or limit of the rain is often a visible part of the picture. Driving along the road I have seen, more than once, rain falling on one side; the ground perfectly dry on the other.

For good liquid measure perhaps, two or more cloud formations occasionally get together and the heavens really let loose. Repeated discharges of lightning rip wide open the covering of black clouds. Deafening thunder roars. The weird, pale darkness is split by brilliant, blinding flashes. Explosions of thunder follow in terrifying volume. The earth trembles. To the tenderfoot, complete destruction, the end of time, seems imminent, especially if the storm comes at night.

At such times the clouds seem to burst. The rain no longer falls in drops. It comes in cupfuls — even bucketfuls. The whole earth is drowned — even the hillsides appear to be covered ankle deep. Gulches and arroyos are filled with raging torrents preceded by a wall of tumbling boulders, brush, and uprooted trees — a roaring, grinding mass of destruction. Often the volume of water and debris is so great that for miles down the arroyo and out on the plain below where no rain has fallen, the drainage channel will be filled with running water too deep for safe crossing. Many a freighter, camped in a dry channel or simply driving his team across it, has lost his entire outfit — wagon, animals, freight, life itself — hours and even a whole day after the storm in the mountains has passed.

Most of these mountain deluges are over quickly. Having discharged their moisture, the clouds soon disappear. Within an hour or two at most, the roads are fairly dry, the sky clear.

When the downpour comes in mid-afternoon, as it usually does, the remainder of the day is pure delight. The fresh-washed atmosphere, the fragrance of fast-growing vegetation make existence joyful. That is the time to drive a span of spirited gray ponies, rarin' to go, hitched to a light buggy with your best girl beside you.

Thanks to the rainfall, the country around White Oaks is fine rangeland, and from the time of settlement the prairies have been dotted with ranches. Before the menace of overstocked ranges blighted the territory, grazing was superb everywhere. Tall grama and red-top grass covered the flats, mountain slopes, and valleys. After the summer rains, the plains were especially attractive in their harvest colors. Almost anywhere wild hay could be cut. There were thousands of acres of it. And there were acres and acres also of wild flowers and blooming cacti.

Wildlife flourished and abounded. Game laws during the eighties and nineties were nonexistent and hunters could find deer, antelope, mountain sheep, bear, mountain lions, wildcats, turkeys, quail, doves, pigeons, rabbits, and coyotes. Occasionally they saw an elk or a spotted cat (jaguar). There were two kinds of deer: the large whitetail and the small blacktail; three kinds of bear: grizzly, brown, and black. The wild turkeys, accustomed to seeing Indians, were comparatively tame. A head shot with a rifle was the approved way of killing them. To use a shotgun was considered bad sportsmanship.

Sharing the range were great herds of antelope, sometimes hundreds in a single band — a wonderful sight. Full of curiosity, they would run along the road within seventy-five yards of a wagon; then stop all at once with heads erect as they stood and stared; finally whirling with one concerted leap to run like the wind. A little later they would repeat the performance, their bushy white tails straight in the air as they ran, a natural target for their enemy, man. They were like a company of well-drilled cavalry, making every movement in unison. After a few shots the main herd might be demoralized and scattered, but the smaller units moved with the same precision.

Hunting antelope was, for the most part, pure wanton slaughter of free, liberty-loving, defenseless animals. Two or three horsemen, dismounting within rifle range of the herd, could shoot until the barrels of the rifles became too hot to handle. First shots might be almost point blank as the gentle animals watched intently these strange invaders of their domain. As the rifles barked, off they would go with a bound, all in one direction. A well-placed bullet aimed at the ground ahead of them would raise a spatter of dust and cause the leaders to turn, the others following in close formation. By repeating these tactics, it was possible to keep the band circling within range. If and when the herd broke in headlong flight, each individual bushy tail made a perfect target, glistening white, the size of a dinner plate — too big to miss. By far the largest percentage of those killed were brought down by *tail-end* shots.

In time their numbers diminished, and then it became "great sport" to hunt them with greyhounds. A number of these nimble dogs, followed by horsemen, would outdistance the fleet but short-breathed animals. A vicious running nip, a broken leg, and the antelope was an easy victim for the killers following on horseback.

A single antelope, dressed out at 75 to 125 pounds and tied behind each hunter's saddle, was brought in as meat — the net product of the so-called "hunt." Carcasses without number and wounded animals hidden in the tall prairie grass were food for coyotes and buzzards.

By 1895 practically all of these beautiful creatures were gone, like the buffalo. The majority of them were *killed for fun.*

The Sun Behind a Cloud

THE SOUTH HOMESTAKE gold mine was supposed to make us rich and White Oaks prosperous. We were all confident of success and counting on the ultimate in profits. We were going to take a trip to Europe, among other things.

For two and a half years Father had been putting work and money into the mill, and now in that summer of 1882 it was about ready for a trial run.

A locomotive engineer from St. Louis named Hodgman designed it and supervised the building. He was willing to *guarantee* a recovery of fifteen percent more than the assay value of the ore.

The plant consisted of a crusher, a bin, a furnace and long, revolving flues for drying the already dry ore, a small arrastra-like iron pan with cast-iron rolls for grinding, and a bell-shaped iron vat into which the discharged pulp was supposed to flow over mercury. The vat had a capacity of 1,000 pounds or more of mercury.

The Big Day approached; everybody was full of excitement. It passed; and nothing had happened. The

26

THE SOUTH HOMESTAKE MILL

mill was a complete and dismal failure. Our hopes were shattered. Our money and time were lost. Our efforts to operate were futile. Plans for a trip to Europe went glimmering. The family was stranded and Father was flat broke, his willingness to trust others exposed. Our disappointment was shared by the entire community.

Once failure had been definitely conceded, the next step was an appraisal of the situation. The consensus was that the South Homestake was down but not out. Having learned their first lesson in mining the hard way, the owners were convinced that the humiliating failure was due to their ignorance and gullibility and was in no way a sign of lack of potential in the mine. While the mill was under construction, development work had exposed a large tonnage of commercial ore and many fine specimens of free gold. To give up and quit was unthinkable. It required three years, however, to undo what had been done wrong and do what should have been done right in the first place.

A new company was organized, Father surrendering the bulk of his interest. Contact was made with Fraser, Chalmers and Company of Chicago, well-known builders of mine and mill machinery. After testing the ores, they contracted to construct a standard twenty-stamp amalgamation mill with a capacity of forty to fifty tons in a twenty-four-hour day.

On the site of the Hodgman failure, utilizing a portion of the old foundation, the new mill was built: gyratory crusher, bin, four batteries of stamps with five 850-pound stamps in each one, inside amalgamation, silver-plated copper plates, and two Frue-Vanner concentrating tables. The latter were not necessary in our operation and were never used. The mill was well built and fully suited to the ores. Recovery was eighty to eighty-five percent of the assay content of the ores.

After the mill was completed, the builder, J. B. Schronz, was retained as superintendent of both mine and mill. Father remained as resident director and manager.

Incidentally, the supply of mercury on hand from the Hodgman mill was sufficient not only to supply this twenty-stamp mill during all its future years of operation but also a good number of others in the district.

This mill, with ordinary and necessary repairs and replacements, continued to operate steadily for nearly ten years and intermittently later until it was practically worn out. It was finally destroyed by fire about 1925.

Pioneer Days in White Oaks

WHEN THE BEAUTIFUL BUBBLE exploded and only wreckage remained, my mother had a decision to make. She decided to stay with Father at White Oaks. She could have returned to her former home, where she had never been without at least one housemaid, and a life of comparative ease, comfort, and protection, but she chose without fear or hesitation to face the future in a log cabin. Her sacrifice was an evidence of character — a tribute to American womanhood.

It is true that New Mexico was good for her. For several months before we left St. Louis her health had been failing. She had "lung trouble," it was said, and was in alarming condition. Her relatives and friends had serious misgivings about her departure (we learned in later years) and were afraid that they were saying final goodbyes. Fortunately, the high altitude and dry climate of New Mexico had a most favorable effect, and it was not long before she regained her normal strength, and though her reserves of energy were never great, she survived for many active years thereafter.

29

Since New Mexico was to be our permanent home, we began to think about improving our way of life. Our furniture in St. Louis had been stored in a warehouse. We made a list of what the log cabin might hold and decided to send for one double bed, a small dresser, two chairs, a sewing machine, and a piano, with a limited number of kitchen utensils and dishes. These things were shipped by rail to Socorro, fifteen miles north of San Antonio, and were hauled thence in a lumber wagon by mule or ox team 105 miles across the rough desert and mountain road. The piano (a Chase) and the sewing machine (a Wilcox-Gibbs) were the first in White Oaks. Incidentally these two articles are still (1945) in the family.

Mother was a talented pianist. She could read and play any ordinary music, not too classical, but she loved best to improvise — just sit down and play, making up the music as she went along. For fun and entertainment she would play "Yankee Doodle" and "Dixie" at the same time, each with a separate hand.

Luckily someone was found to tune the piano. Musicales followed, attended by whoever might care to come — and there were plenty. Those young men were mostly miners and apparently roughnecks, yet a surprising number were college graduates and from cultured families. And how they loved to sing! Mother sent for half a dozen or more gospel hymn books and an equal number of collections of college songs. The boys sent for some special things themselves. Soon other instruments appeared: a violin, a flute, a guitar, a banjo. But the piano was always in the lead. Other activities developed with the singing. We had recitations, readings, talks, even a "Literary Club" — at that time the rage.

SCALE IN FEET
0   500   1,000

ROCKY PEAK

WHITE OAKS

15 & 19   16

JANE'S PLACER

WHITE OAKS TOWNSITE

1. William Watson home
2. Sidney Parker home
3. John Y. Hewitt home
4. Susan McSween Barber home
5. Charlie Littell home
6. Albert Ziegler home
7. Matt Hoyle home — built 1893
8. W. C. McDonald home
9. Joseph Grieshaber home
10. Erasmus W. Parker home
11. Dr. M. G. Paden home
12. Paden Drugstore
13. Gumm Lumber & Building
14. White Oaks schoolhouse — built 1894
15. Paul Mayer Livery Stable
16. Jane Gallacher home
17. Ozanne Hotel
18. A. N. Price home
19. Ziegler store
20. Little Casino Saloon
21. Young & Taliaferro store
22. Dr. Alexander G. Lane home and office
23. Lois McDonald home
24. Gumm home
25. J. P. C. Langston home
26. Charlie Mayer home
27. Stewart home
28. D. L. Jackson home
29. Exchange Bank Bldg.
30. Charlie Mayer's Blacksmith Shop

Don Bufkin

Long before the piano arrived, Mother had organized
a Sunday school using lessons printed in advance and
received by mail from a missionary society in Boston.
Despite, or perhaps because of, the environment, she
was determined that her children and the other children
of White Oaks should not be deprived of religious and
moral influences.

The Sunday school was followed by an elementary
day school for the children, a room equipped with desks
and benches in Dr. A. G. Lane's already crowded residence. Dr. Lane himself was the teacher. He was a kindly,
dignified Southern gentleman with a flowing gray beard,
sincere in his desire to aid the town and its children.

The need for a public building, a town hall, led to
the construction of a 24- x 48-foot building with walls of
4 x 6 planks laid flat and spiked, and a corrugated sheet-
iron roof. Most of the material was furnished at cost, or
less. Some of the labor was donated. What money was
needed was raised by subscription.

When this building became available, the question of
a place to hold church services was solved. Because of its
liberal doctrine, the Congregational organization was
chosen. Twenty charter members belonging to nine church
groups worked together. They purchased a small melo-
deon (foot-organ) to provide music for church and Sun-
day-school services.

The town hall became the center of all public gather-
ings: church, Sunday school, day school, dances, political
meetings, home-talent theatricals, children's entertainments, and so forth.

The Reverend Lyman Hood, moderator of the Congregational Church and its missions in New Mexico, held

the first church services when he came in 1884 to organize
the church. With his assistance (and later with the help
of the Reverend E. H. Ashman, his successor), the Home
Missionary Society of Boston supplied the local pulpit.
As I recall it, the minister was paid $75.00 a month. One
third of his salary came from the Missionary Society.
The rest, plus heat, light, janitor service, etc., was raised
by church members, merchants, and other townspeople.

All these civic improvements were important and
necessary, but just as important to us were the improvements we made in our living arrangements. When Mother
made the decision to remain in White Oaks, one of the
first requirements was a fence around the log cabin. The
post-hole digger and carpenter's helper, incidentally, was
George L. Ulrich, about twenty years old at the time.
Later he became one of the prominent men of the community, vice president of banks in White Oaks and Carrizozo, still later a close friend of Governor W. C.
McDonald and deeply involved in state affairs in Santa Fe.

At the rear of the enclosure we built a stable, cow
shed, and chicken coop. In no time at all we boys learned
to feed and care for the animals. We curried, harnessed,
and saddled the horse, milked the cow, and fed the chickens. When the cow went dry, a couple of goats solved the
problem — and we learned to like goat's milk.

Jamie (later we called him Jim) because of his
advanced age was the director and chief counsellor. He
was two years older than Frank and three years older
than I. Frank and I, eleven months apart, worked and
played together, though without any marked exclusive-
ness. Frank was the best worker. I was always the "willing
helper." Mother, for reasons of necessity but very wisely,

directed her efforts toward training her boys to help and to do the housework. This practical education did us no harm and often came in handy in later years.

By "housework" I mean just what the word implies: washing dishes, sweeping, dusting, churning milk to make butter and cottage cheese, killing and dressing chickens, slicing ham and bacon, and *cooking*. We had our daily cereals, meat, and vegetables. In addition we baked half a dozen loaves of "raised" bread once a week — also a pan of raised biscuits — with baking-powder bread and biscuits between times. Pies, cakes, and cookies were part of our usual fare; never enough.

Somehow, maybe because of unmerited but accepted praise from others, particularly my two good brothers, the honorable title of Chief Chef was attached to my kitchen apron — a title which, in time, became an asset.

Then there was the family laundry. No hot or cold running water, no electric washing, rinsing, ironing machines in those days. It was a case of backache and arm-strong equipment only. We heated water on a wood-burning stove in a deep, oblong "boiler"; transferred it by means of a galvanized bucket. The rest of the equipment consisted of a wash tub and corrugated washboard, two rinsing tubs, and a hand wringer. The clotheslines were in the back yard. The dry clothes were sprinkled and ironed — three or four heavy, solid, cast irons, heated on the kitchen stove, being used in rotation. Not my idea of work for a lady — then or now.

We felt grateful to be relieved of the washing and ironing, for a time at least, by Mrs. Madden who with her Irish husband Pat lived across the arroyo. She was the only washerwoman in town during the early days and ours was the only wash she would condescend to do. We always had a kindly feeling for our friend Mrs. Madden.

The floor scrubbing continued to be our responsibility. As regularly as Saturday morning came each week, we three boys got down on our hands and knees and scrubbed the kitchen floor, one step below that of the log cabin. It had twenty-two boards, a number not divisible by three. This often meant an argument, possibly the start of a free-for-all which had to be settled by Mother, perched on a stool in the doorway. Each boy had his own dishpan, hot water, scrubbing brush, washing and drying cloths; and woe to him who, inadvertently or otherwise, slopped water beyond his allotted crack. As umpire, Mother saw to it that her decisions *stood*.

It was not all work, of course. We had our diversions. Frank and I had a young calf which we "broke" so we could ride him bareback around the yard and the neighborhood. One day when Mother was not home, we rode him into and through the house, around the center posts of each room, and so on out. When Mother came home, she knew something had happened. Frank and I straightened out the mess.

And our calf! It was a sad day when the family decided he was grown and should be sold. The butcher paid us twelve dollars for him. We took turns riding him to the slaughterhouse on the other side of town. Neither Frank nor I ever forgot that experience as, with tears streaming, we left our pet calf and walked home without him.

Community set up own Church & School? Cultural activity.

Early Days – 1880-1885

GOOD OLD DAYS, days of trial and change, of give and take, when friendship was deep and sincere, when a man's word was as good as his signature, when bad men who shot out of turn were annihilated, not always by due process of law.

He who has never traveled cross country in a Concord coach alongside the driver, high on the seat above the boot, has missed a lot of scenery and some wonderful tales of Wild West adventures. If he has never slept inside on top of big sacks filled with U. S. mail, he has missed an incomparable sleeping experience.

The early and rapid growth of White Oaks, as previously mentioned, called for a stage line to cover the ninety miles between San Antonio and White Oaks. At first bi-weekly, the service soon was on a daily schedule. At the river crossing near San Antonio, passengers and cargo were transferred by ferry during high water. During the dry season, the stream was fordable.

Shortly after the Santa Fe reached El Paso (where it made connection with the Southern Pacific) in 1881, a branch line was built ten miles east of the Rio Grande to promising coal fields at Carthage. A bridge was constructed for the river crossing. Carthage thereafter was the White Oaks stage-road terminus.

Standard Concord coaches were used by the stage line, four horses to the coach. The body of the vehicle was enclosed with doors on either side. The high driver's seat and the boot (deep space for cargo) were outside. An extended rack for trunks was at the rear and a railing around the top took care of extra baggage. The whole was suspended between the axles on heavy leather "sway" springs. The vehicle in motion felt more like a cradle than a spring wagon ever could. The coaches were made at Concord, New Hampshire. They weighed 2,400 pounds and cost $2,400.00 — a dollar a pound. They were the toughest things on wheels — never broke down, and it took years to wear them out.

The stage road, entering White Oaks from the west, continued through town, thence northeast and north to Las Vegas. It was the sole highway across the plateau-valley on which the town was situated.

The nearest town was Lincoln, the county seat, forty miles by road east. The population was about 300. It was a distribution center for the surrounding section, mostly cattle and sheep ranches. To get there from White Oaks, it was necessary to go west past the Hightower ranch to the western entrance to the valley, turn south, then east around Carrizo Mountain and across the southern portion of Carrizozo Flats, thence eastward across Nogal Divide, on past Fort Stanton, which was ten miles west of Lincoln.

Lincoln had recently (1880-1881) become famous by

THE EARLIEST STAGECOACH

*John M. Decort, driver,
at Red Canyon in the
Oscura Mountains on his
way to Carthage (1888-89)*

— *John Kelt and Frances Ozanne Lanier*

reason of the "Lincoln County War," a cattlemen's dispute with a difference — the difference being the terrorism created by twenty-one-year-old Billy the Kid, New Mexico's number-one bad man. Father had met and talked with the Kid. We all knew Pat Garrett.

Fort Stanton was our nearest telegraph station until 1901. It was a United States military post then. Later a Navy hospital for tubercular cases was located there, and during World War II it became an internment camp for German prisoners.

Shortly after the San Antonio-White Oaks stage-mail route was established (1879-1880), an auxiliary route was added which extended to Lincoln. From there the road continued to Roswell on the banks of the Pecos River, sixty miles east.

These were the embryonic days of White Oaks, and

they were no doubt full of fun, tragedy, and disappointment — local history in the making. Between the ages of eleven and fourteen, however, I was too young to be greatly impressed by such matters. Nevertheless the mixed character and culture of the residents, both mobile and fixed, was evident. They ranged from ignorant nomads — cowboys and prospectors — to college graduates. They included good men and bad, gold-hungry adventurers and people who just came to look around. Horace Greeley's widely publicized advice — "Go West, young man, go West!" — was the motivating force for the latter group.

A high percentage of the money in circulation was brought in by the new arrivals. Without them we would have had to get along on trust and faith. Conditions, I might add, required a lot of faith and optimism. The

Governor W. C. McDonald

Dr. Alexander G. Lane,
pioneer physician

BUILDERS OF WHITE OAKS

Opposite page — back row: *Attorneys William Watson
and John Y. Hewitt; Watson Hoyle, who built the
Hoyle house.* Front: *Attorney Harvey B. Fergusson
between two unidentified men*

*Gus Wingfield (left), house builder for the
White Oaks Building Company, and Dr. M. G. Paden*

narrow, rich channel of placer gravel in Baxter Gulch, a short half mile in length, had been quickly exhausted and turned over to the more patient and less avaricious Mexicans. So far, only one rich outcrop of gold had been uncovered, and no real production was achieved there until 1885 — *six years after discovery.* Money spent on the South Homestake went mostly for equipment and freight. The local payroll was a minor item. The little money in circulation, brought in by newcomers or paid out for labor, was raked in and carried away by itinerant gamblers, while struggling merchants with limited capital had difficulty in replacing their original stock of merchandise. Naturally a slump occurred with the failure of the Hodgman mill in 1882, and times were harder than ever.

This long period of non-production, however, may have been a good thing in disguise. During this transitional period many undesirables left town. The worthwhile citizens remained, hoping for increased work opportunities at the mill and taking comfort in the possibility of other discoveries.

Intelligent prospecting took the place of grab-and-stake methods, but for some time most of the traffic in claims was the result of transfer rather than discovery. Disillusioned by the absence of rich outcrops or by the lack of prospective suckers, many locators failed to do their assessment work. The government required the owner to do one hundred dollars' worth of labor each year or forfeit his claim. Midnight, December 31, was the deadline. Bonfires built by "jumpers" dotted the hills surrounding the town. These were the first comers. Others arrived a little later. Often there would be two or more claimants for the same abandoned ground. A compromise was sometimes effected, but usually the most belligerent gained possession. Rifle or six-shooter shots might indicate a duel, but so far as I know, no one was ever hurt in such a mixup around White Oaks.

By 1885 White Oaks had assumed the semblance of a real town. Streets and cross streets had been surveyed and laid out. Store buildings and dwellings were all in line. The early-day boom-town atmosphere, featuring saloons, wide-open gambling, and guns and Bowie knives dangling from belts, had practically disappeared. The majority of the citizens were law-abiding, progressive people, unafraid and confident of the future. M. H. Bellamy, the justice of the peace, represented law and order. He was assisted when necessary by Sheriff Pat Garrett and J. P. C. Langston, the constable. There were at least

Albert Ziegler

MERCHANTS OF WHITE OAKS

Mr. and Mrs. L. W. Stewart

Notable citizens of White Oak

three young lawyers: John Y. Hewitt,[1] H. B. Fergusson[2] (later U. S. Senator) and George Barber.[3] William C. McDonald,[4] first governor of New Mexico after statehood, was U. S. Deputy Mineral and Land Surveyor. M. G. Paden and A. G. Lane were our two doctors; each one had a drug store.

Local, Territorial, and national news and gossip were supplied by a weekly newspaper, *The Golden Era*, renamed in after years *The Old Abe Eagle*, and still later *The White Oaks Eagle*. It was edited by Major William Caffrey, a typical Wild West word-shooter whose lethal and horrendous phrase-making would make present-day Hollywood oratory seem gentle and political discussions tame. On the staff of *The Golden Era*, as reporter, type-setter, and devil, Emerson Hough developed his wings. Later he became a famous and popular novelist — a con-tributor to the *Saturday Evening Post*. His book *Heart's Desire*,[5] written long after he left, is centered on White

---

[1] John Y. Hewitt (1836-1932), born in Ohio, served as a Union soldier 1861-1865, came to White Oaks in 1880 and in his time played many parts: lawyer, mine owner (the Old Abe), newspaper editor, and first citizen of White Oaks.

[2] Harvey Butler Fergusson (1848-1915) was an Alabama boy who became a lawyer and came to New Mexico to take part in a lawsuit over mining titles at White Oaks. He remained to become a political leader, serving as a Democratic member of the U.S. Congress, and a noted trial lawyer. His children Harvey and Erna gained distinction as writers.

[3] George Barber owned the Three Rivers Ranch from 1885 to 1889 but is best known as the second husband of Susan McSween, her first dying in the Lincoln County War.

[4] William C. McDonald (1858-1918), a New York Stater, came to New Mexico in 1880. Among other enterprises he man-aged the Carrizozo Ranch. Morris Parker's mother-in-law became his wife.

[5] *Heart's Desire* (Hough's name for White Oaks) was pub-lished in 1903. A product of Iowa, Hough came to White Oaks in 1883 to practice law but did very little legal work and left in 1884.

Oaks. Although he builds the railroad and organizes the town of Cloudcroft twenty years too soon, many of his places and characters are familiar to the old-timers. Some of the latter are given their right names.

White Oaks at no time had any public utilities: no sewer, gas, water mains, telephones, or electricity. The only sidewalks were in front of the store buildings, and they were made of rough planks.

The stores were all on the main street, White Oaks Avenue. General-merchandise establishments (groceries, clothing, hardware) were operated by Jones and Taliaferro, W. H. Weed, A. Ridgeway, A. L. Stewart. Clothing stores were run by A. Ziegler, A. Whiteman, Sol Wiener. T. B. McCourt sold tin and sheet-metal; Starr Page handled jewelry; Mr. Corey repaired shoes; Paul Mayer owned the livery and feed business; Charlie Mayer and Joe Biggs were the blacksmiths. Dr. M. G. Paden, Dr. A. G. Lane, and J. A. Reid dispensed drugs. We had a Chinese laundry (belonging to Mr. Chew), a barbershop, and two hotels. Urbain Ozanne's was on the main street; the other was a block away near the Congregational church and the town hall. Most of these businesses were established before 1885. The bank and its building came later — 1893.

The town never had a red-light district or any noticeable number of wild women. Saloons came and went, but only one remained year after year, for months at a time the only one in town. Gus Shinsing was the proprietor. With his unassuming wife he received the merited respect of the townspeople. Gus himself said, and no doubt it was true, that his business was principally in beer and wine. There was very little profit in hard liquor because the leading merchant, W. H. Weed, pretty much monopolized the retail sale of whiskey. In a side room of his

[handwritten marginal notes: White Oak, proprietor, Drug Store, Chinese, laundry, Jewish, Shopkeeper]

store was a barrel of rye and another of bourbon resting
on a platform, the faucets at a convenient height, jigger
glasses available. Miners and teamsters, cowmen and
peace officers, walked through the store into this room
and helped themselves, no one looking on. When leaving,
each one gave a finger signal to a clerk and he made the
proper entry in the customer's account.

An occasional snitch might occur, but evidently the
profit was satisfactory. And there were obvious advantages both to the drinker and to the community. There
was less lining up against a bar, less treating, less exces-
sive drinking.

For those who drank water, there were wells. A few
residents had their own, dug down to water level at
thirty-five to forty feet below the surface. The water was
"hard," impregnated with alkaline salts. The public was
dependent for its drinking water on a water wagon, a
cylindrical wooden tank hauled by a span of horses. Delivery was made to a barrel standing in the yard of each
dwelling. The price was fifty cents, later reduced to
twenty-five cents, a barrel — forty gallons.

In later years a number of the better houses, those
with pitched roofs of sheet iron or shingles, had cisterns
for catching rain water, soft water used only for drinking
and bathing. To soften the well water, we used the root
and core of the *amole* cactus, abundant in the vicinity.
We shredded it, pounded it, and stirred it into the water.
This native Mexican stunt later became the basis of the
process which produced Amole soap.

The collars of dug wells were planked over. A three-
foot-square box at the center supported a frame which
carried a twelve-inch sheave pulley, and through the
pulley ran a rope with a bucket at each end. A hand-over-

hand pull raised the loaded bucket, at the same time lowering the empty one.

We certainly did not live a life of luxury, but we ate surprisingly well. In addition to the bread from our weekly baking we used a great deal of cornbread, cornmeal mush, and oatmeal, along with hot cakes drenched in molasses or syrup. Our coffee was the well-known Arbuckle's roasted, purchased in one-pound packages at three for a dollar. We ground it at home in a grinder fastened to the kitchen wall.

People supplied fresh produce from gardens

Fresh vegetables, in season, came from one local farmer, a Frenchman of peculiar character. He spoke but little English and his prices were sometimes confusing: One bunch of onions, ten cents, two for a quarter.

Tom Osborn, a long-range cross-country peddler, provided our main supply of oranges, apples, watermelons, and so forth, and our potatoes, onions, and turnips on occasion. Back and forth he traveled with his team and covered wagon — from Las Vegas 175 miles north; from Tularosa 45 miles south; from Hondo and Ruidoso 60 miles east. He was one of the most popular and welcome tradesmen in our midst. Two, three, or more times each year, he came in loaded and went out empty with his profits, leaving us happy with the much-appreciated luxuries he had provided.

Sugar came in hundred-pound sacks, flour and beans in fifty-pound sacks, rice, salt, and other commodities in smaller sacks. We got our salted mackerel in wooden tubs. Codfish came in slabs, bacon in sides, ham and shoulders as the Lord made them. Canned goods — corn, tomatoes, peas, condensed milk when it became available — were often purchased in case lots and shipped in with other freight.

THE SOUTH HOMESTAKE COAL MINE

— John Kelt

Kerosene (coal oil) lamps lit the dwellings. Tallow or wax candles were used for underground work in the mines.

Wood for fuel was plentiful on the nearby hills and cost $2.00 a cord delivered until the price went up to $4.00. Cutting to stove length and splitting cost extra. Most people cut their own. Timber and lumber, delivered, cost ten to twelve dollars a thousand board feet. By 1892, these prices had increased about fifty percent.

An especially useful asset was a hard, semi-anthracite coal of extra quality which burned with very little ash. The South Homestake (Father's company) owned the only operating coal mine prior to 1892, a seam four feet thick with slight dip into the hill, situated about two and

a half miles east of town just beyond White Oaks spring. It was easily accessible and easy to work. For years the coal was mined on contract and delivered to surface bins at seventy-five cents per ton. Hauling to town cost sixty cents per ton. The price of the coal delivered at our house was $1.35 per ton; $2.25 at the mine; around town, $3.00. Fireplaces and grates were commonly equipped with grates which would burn both coal and wood. The superior quality of the coal made it usable for all purposes: heat, steam, coke, blacksmithing. When a few years later we moved to El Paso, we paid $9.00 a ton for a vastly inferior grade of sub-bituminous coal which left three times the ash of the White Oaks coal. It was a bitter shock to us.

Wages in those early days were about the same as

miners wages (margin handwriting)

in other Western mining camps. Miners and timbermen were paid $3.00 a day; muckers (shovelers) and car-men, $2.50; common laborers, $2.00; mill amalgamator men and battery men, $3.50; blacksmiths, $3.00 to $3.50; master mechanics, mine foremen and shift bosses, $3.50 to $5.00.

10 hr days Sunday for clean up (margin handwriting)

Day shifts at the mine were ten hours long, night shifts nine hours, six days a week. Mill shifts were twelve hours seven days a week. Sunday was usually the hardest day of all, since it was reserved for cleanup and repair jobs. Sunday night was the only regular time off. Overtime, for necessary repairs, and so forth, was paid for at a standard rate.

Many a day in later years, when we boys worked in the mill, have I walked a mile each way, going and coming, with lighted lantern in one hand and lunch box in the other. The winter days were so short that we began and ended them in darkness, but it was all taken in stride, nothing to kick about. The miners had but little advantage over us boys, getting to work at seven, quitting at five-thirty with a mile-and-a-half walk each way.

Work and Play – Gold and Money

WHEN "DOC" REID, a druggist and pharmacist by training, first came to White Oaks, he helped with the construction of the Hodgman mill. When that failed, he was kept on as a watchman, and also served as the company assayer. Perhaps because I was "willing," he taught me the first principles of the latter art: to crush and grind samples and to make cupels. He thus unwittingly started something that would be useful to me in later years.

Fond of children, he built for us boys a small, strong wagon. The wheels were about twelve inches in diameter, the body forty inches long, the width about twenty inches. It had iron axles, braces, and supports. We already had a burro, and it did not take long to break him to harness; ditto the calf mentioned above and a goat which we then had for milk when the cow went dry. With one of these for motive power, errands around town, trips to and from the stores, and the transportation of cordwood for the family became a source of pleasure to us rather than work.

The placer deposits of Baxter Gulch were about three-quarters of a mile from where we lived. We boys had watched the miners, mostly Mexicans, digging and washing the gravel to extract the gold. In due time we conceived the idea of doing likewise. With our wagon and burro, a ten-gallon keg for water, a pick and a shovel we started our pursuit of fortune. The keg filled with water was pretty heavy for the wagon and tended to lean to one side or the other, but with one of us leading, riding, or driving the burro and the other walking alongside to balance the keg, we managed to deliver a fair portion of the load.

What with housework and chores, we could toil at the placers only at odd times, perhaps three or four hours a day — and that was enough since the work was hard. Several days of labor were necessary to sink a hole six feet deep in order to reach "pay dirt" near "bedrock," where the gold occurred. When the depth became too great for us to throw the gravel out with a shovel, one boy would work in the hole filling a two-gallon bucket while the other stood on the brink and hauled it to the surface with a rope.

To extract the gold we had a regulation "cradle" or "rocker," also made by our friend Doc Reid. In addition we had a standard "gold pan" made of sheet iron, fourteen inches in diameter and four inches deep. During the summer of 1883 we worked like Trojans, adding little by little to our prized production in spite of difficulty and discouragement while the smiling Mexicans encouraged us and cheered us on.

The placer gulch was only about half a mile long and had been worked for several years, yet there were still groups of Mexicans at work on left-over portions.

These thieving pirates had some good laughs at our ignorance and were the main cause of our quitting when we did. After we had spent three or four days digging a hole to pay dirt and were all set for a good clean-up next day, we would find that the good pay gravel had been removed overnight — our work lost along with the gold.

We did not lose it all, but what we did get cost us enough. The work was really hard as we dug the deeper holes and washed the gravel in the rocker. For "safety" Jamie elected himself custodian of the small glass phial in which our accumulating gold was kept. The merits of this procedure were discussed for years afterward. The summer's work produced nearly one ounce valued at $18.00. And were we proud youngsters? Yes!

What to do with our wealth was the big problem. On a trip to St. Louis the following winter, Father took the gold and had a heavy band ring made from it. The ring was intended for Mother, but it was too big and heavy for her and Father kept it. He wore it on his little finger as long as he lived. Frank's widow, Bruce, now has it — a prized heirloom of early days.

Our placer mining came to an end when our wagon was demolished by a runaway burro. With no consideration whatsoever for the boys (who adored him) or for the wagon, off he tore — across deep gullies, over big rocks, through heavy brush where the wagon had no chance at all. The destruction was complete. Bystanders who witnessed the runaway expressed amazement that Frank and I, trying to hold on, escaped being a part of the finished wreck.

Thenceforward we attempted Bigger Things, enthusiastically making play out of activities which older persons call work. Father owned two portable sawmills out among the pines ten or twelve miles east of town. During the summer of '84 one of the mills was idle. Two of the big log wagons and several yoke of oxen were brought in, the wagons were stationed outside our yard, and we boys were detailed to look after the oxen. They were turned loose for grazing and it was our job two or three times a week to look for them, drive them back towards home, and see that they did not stray and get lost.

It was not long until Frank and I had learned to put on the heavy yokes, hitch the teams to the wagon, and drive around. Our driving equipment consisted of a long sharp stick and a long braided-leather "rattlesnake" whip which we soon learned to crack like any old bull skinner of that period. We also learned the "Gee-Haw" language.

The slow gentle animals seemed to sense that we were a couple of young inexperienced youths, and they acted accordingly. I still contend that back of those big round eyes there lurked a degree of potent humor which could account for the number of mixups we got into.

Two oxen were required to move the wagon on a level road. With any load at all, at least two yoke were necessary. For our purposes we needed three yoke.

After a few practice trips, we went into the wood business, selling stove wood to people around town. We charged two dollars a cord for good, dry juniper, oak, cedar, and piñon — two dollars more to cut it into cookstove lengths and split it. Great fun!

Then one day we had a mixup, somewhat worse than usual. We were coming out of the timber, going downhill with a full load of wood, when one of our Gee's or Haw's failed to register. The hub of one of the front wheels ran into a tree, causing a quick stop which threw the oxen sideways. It is a wonder their necks were not broken. These wagons were built so big and strong that nothing would give way. The weakest point of the entire outfit was the ox's neck which was fastened to the yoke by a curved piece of wood called the bow. A camel can go through the eye of a needle about as easily as an ox can get rid of his bow without help.

The sudden jolt caused a scramble and an awful mixup.

Our problem now was how to get going. It was impossible to back up. We finally decided to chop the tree down, cutting below the hub. This was no small undertaking since the tree was seven or eight inches in diameter. We got it down, however, but we forgot to set the brake or tell the oxen when it was about to fall. At the sudden release, the wagon shot forward, downhill, directly into the quietly dozing beasts, bringing on the final and ultimate degree of confusion.

We got her stopped somehow, but the result was anything but pleasant. Several of the oxen were badly bruised and hurt.

Thus ended our bull-whacking career as well as our wood-supply business. Father made us quit.

I was not long among the unemployed. One of the sawmills had been moved to Nogal Gulch, fifteen miles out on the road to Lincoln. The foreman came to town looking for a camp cook to replace one who had given up and gone away. On hearing his complaint to Father about the difficulty of finding a man, I volunteered to take the job. Father and Mother both demurred, but after considerable argument they consented to let me go. I was excited, scared, doubtful of my ability to do the work, but not enough so to back down.

The crew consisted of ten or twelve men, a fair-sized lot for any cook. They treated me royally, however. They helped me; they never complained; they even praised my cooking. I gave them hot cakes, bacon, eggs, steaks, roasts, bread, pies — everything the larder supplied. It was pretty hard work, however, and soon became monotonous and tiresome. The hours were long. I got up at 4:30 and was busy all day. For "recreation," during spare moments in the afternoon, I learned to use a yardstick for measuring the length and diameter of rough logs — then to compute the number of board feet each contained before it was converted into lumber.

At the end of six weeks Mother, fearing the results of prolonged contact with lumberjacks and other rough characters on her boy, sent for me to come home. Bill Taylor, our "handy man," came over in Father's two-wheeled gig behind Polly the mare, bringing a cook to take my place. I was willing enough to go. Home looked good to me and I had $37.50 in my pocket.

On the road we passed two big loads of wild hay cut from the prairie. Recognizing the Mexican drivers, we stopped to talk. They told us they were taking the hay to White Oaks to sell. The price was twenty dollars a ton — forty dollars for the two loads.

I could feel the cash money in my pocket getting hot. We began to dicker, and presently we had made a deal whereby they were to unload the two wagons in our back yard, stack the hay properly, cover it with a tarpaulin, and help build a fence around it — all for $37.50.

Before I left White Oaks for school a couple of weeks later, I had arranged with Bill White, the contractor who hauled ore from the South Homestake mine, that on or about Thanksgiving Day he would buy the stack "as was" for $60.00 — which he did, paying the money to Father for my account.

Thus in that year I held my first real job, made my first business deal, collected my first profit. Conclusion: making money was easy. Not until later years did I learn that *holding on* was the hardest part.

During these boyhood years, Mother was our school-teacher —a trying and difficult task for one who had had no previous experience. From St. Louis she obtained text-books with regulation contents so we could continue in our grades. At times she appealed to Father for help, but the results were not encouraging. Our frequent failure to grasp his explanations frustrated his efforts and strained his patience.

Because we had our own textbooks and because there was some uncertainty about what could be expected of the town (public) school, we preferred our family school and considered our system more efficient. During the school season we spent regular days and hours in study and recitation.

It was not long until the three of us, being so near the same age, were studying the same lessons. This, in a measure, simplified the teaching. It so happened that I was good at mathematics. Mother afterwards told me how she managed to keep me a page or two ahead of the others in order to lessen her difficulties in solving difficult problems.

I was chosen to be the first to leave home for school. Time: fall, 1885; age: fourteen; place: Penn Yan, New York, Mother's old hometown. On consulting the curriculum of the academy where I was going, we found that I would need half a semester of geometry to enter the grade for which I was otherwise prepared. A few weeks under the tutelage of Mr. W. C. McDonald, the local surveyor, remedied the defect. The teacher even bragged about his pupil. W. C. McDonald was afterwards the governor of the state.

At the time I left, Frank and I each owned a burro. Several weeks after my departure, Frank traded the two for one — a larger and better animal. It was customary, when he was not in use, to turn him loose at night outside the home pasture with a bell attached to a leather strap around his neck.

One day the new burro, a big white animal, was missing. After searching for several days, Frank found him lying in a ditch with one hind leg caught under the bell strap. Frank cut the strap, but the burro was too weak to stand. Frank hurried home and brought water, hay, and grain. The burro recovered, and a short time later Frank sold him for $15.00.

I always contended that half that burro belonged to me — hence, half the money. Frank's argument was that since I was in New York and the burro was lying in a ditch in New Mexico with one hind leg under the bell strap, he (the burro) would surely have died had not he (Frank) saved his life. Hence the money was all his. The argument continued for years; the status quo remained unchanged.

Gold in Production

NOBODY EVER DOUBTED that there was "gold in them thar hills." Others besides ourselves came, saw, and decided to try their luck. By 1885 we had four mills in place of one. As time went on, many new facilities were provided, including cheap lumber and timber, plenty of labor, and good freighting facilities. In other words, preliminary expenditures of time and money were reduced, and any new venture looked easy. With experienced supervision, success might not have been difficult — but where was the experienced supervision to come from?

The Little Mack-Henry Clay properties, about a mile north of the Homestakes, showed a strong true fissure vein with exposure of specimen gold ore. The New Jersey and Philadelphia capitalists who acquired these properties had about the same knowledge of mining as the owners of the South Homestake. By accepting the same kind of friendly, inexperienced advice, they too spent a considerable sum of money building a mill designed to save more gold than the ores contained. The result

matched exactly what happened to the first South Homestake mill. The method of grinding was as bad, or worse,
using twelve-inch cast-iron or steel balls inside an iron
pan about five feet in diameter and thirty inches deep.
The pan was made to revolve with the expectation that
the balls would roll and thus pulverize the ore. They
did no such thing, and the mill was the second in White
Oaks to fail completely. White Oaks ores took a lot of
punishment before any real production started.

About this time an experienced mill man named
Captain C. H. Gallagher, sensing the need for a custom
mill, brought in and installed a second-hand standard-
type Colorado-California ten-stamp amalgamation mill
well suited to treat the free-milling ores of this district.
For several weeks it was kept busy on Little Mack-
Henry Clay ores. Then all the available commercial ore
was exhausted. Exploration failed to develop new ore
bodies. The properties closed down; the company dis-
banded. Several times in later years the mines were
refinanced and new attempts were made to find more
ore. They did not succeed. The Gallagher mill remained
idle until the discovery of the Old Abe mine some six
years later, in 1890. Until that year White Oaks had
one (good) mill too many.

Meanwhile, in the fall of 1884 and the spring of
1885, the half claim (750 feet by 600 feet) known as
the North Homestake was making history. Quietly and
without fanfare it was purchased by James A. Sigafus of
Colorado, the one and only experienced mine operator
to enter the White Oaks field. Frank Lloyd was his
manager.

Exploration and development received their first
attention. Using a hand windlass, they sank a shaft to

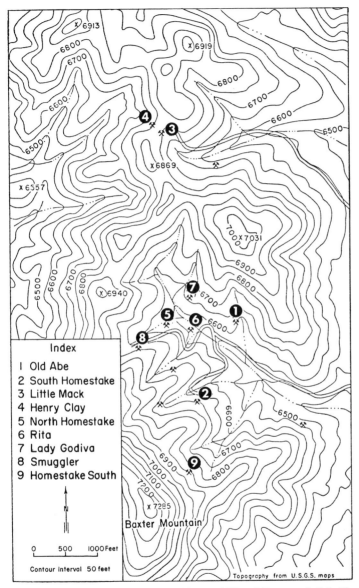

Index

1 Old Abe
2 South Homestake
3 Little Mack
4 Henry Clay
5 North Homestake
6 Rita
7 Lady Godiva
8 Smuggler
9 Homestake South

N

0   500   1000 Feet

Contour interval 50 feet

Topography from U.S.G.S. maps

Baxter Mountain

— *New Mexico State Bureau of Mines and Mineral Resources Bulletin 67*

PRINCIPAL MINES IN THE WHITE OAKS DISTRICT

the depth of 100 feet with drifts at the fifty-foot and hundred-foot levels. The shaft was then equipped with a fifty-horsepower mine hoist.

A mill followed shortly. It had a crusher and two five-foot Huntingtons (copper amalgamation plates). Its capacity was thirty to thirty-five tons daily (twenty-four hours). It was situated on the same hill slope as the South Homestake, some 500 yards distant. A well near the mill provided water. The operations were successful from the start — practical, conservative, slow moving, profitable.

It was characteristic of the management that the company residences were built near the mill so the men could give close attention to their work. As a result, they displayed only slight interest in local affairs, but they served the ends in view: low costs, steady production, and *profit*.

Mr. Sigafus was content to let Mr. Lloyd — a most capable man — run the mine his own way. One reason was the fact that he already had plenty of money. Just before coming to White Oaks, he sold his one-third interest in the Robert E. Lee mine in Colorado for three million dollars. He had a partner in the White Oaks venture, a Dr. Willard Hunter of Louisville, Kentucky, who never, so far as I can remember, appeared in White Oaks, and Mr. Sigafus visited the mine very infrequently, but since things went very well without him, he probably saw no need for active supervision.

Following a line down the center of the 750-foot length of the North Homestake is a perpendicular true-fissure vein, the outcrop ill defined on account of the soil covering. Therein were two separate ore shoots about 400 feet apart, neither of which was encountered until some time after the original discovery and location of the claim. Two shafts were sunk: the first, the main shaft, to a depth of 1,066 feet; the second to a depth of 500 feet.

The two Homestakes went into production about the same time in early 1885.

For three years the South Homestake maintained steady production on account of a glory hole, an open cut known locally as the Devil's Kitchen. It was about 100 feet square and 75 feet deep. About 50,000 tons from this area were mined and milled.

Because the bottom of the Kitchen contained too much rock and insufficient gold, Father released Mr. Schronz, a good mill man but limited in his qualifications for underground mining. E. W. (Father's local and familiar appellation) assumed the management with Joe Grieshaber as mine foreman.

The gold values near the surface were distributed in the decomposed, disintegrated diorite rock formation; those in depth were confined to narrow quartz veins between hard walls which contained no values. A U-shaped tunnel was driven into the hill with two surface entrances and the turn immediately under the glory hole. At that point a room was excavated with space for a hoist, sheave, compressor, and other equipment. A ventilator hole brought in fresh air from above. From this room a two-compartment shaft was sunk to a depth of 600 feet with levels at fifty- and seventy-five-foot intervals.

All this equipment, as well as the shaft timbers, was destroyed by an underground mine fire on July 1, 1891, in the hoist room at the collar of the shaft. It was started, presumably, by a lighted candle left burning on an oil-soaked shelf. Only two men were on shift, both of whom lost their lives.

The fire was first noted early in the morning by men leaving the night shift at the North Homestake. The whole town was aroused, and it required the continued work of several groups of workmen for two days and nights before the bodies were brought to the surface. Evidently the men were eating lunch at midnight. Lying beside their open lunch buckets, the bodies were found in a natural reclining position, as though they had simply closed their eyes and gone to sleep. They were black as Negroes, literally baked. The skin, gently touched, would slip and peel from the flesh, like paper on greased glass.

The entire installation was rendered useless. Better to start anew than to attempt rebuilding of the old; so a new shaft was sunk from a point higher on the hill above the Devil's Kitchen, the deeper levels connecting with those of the former workings. Tonnage and values below the depth already attained proved erratic and disappointing. Hence no new exploration was scheduled and no further depth was reached.

Meanwhile another subdivision of the mine known as the North Shaft was producing good ore, in a measure offsetting the disastrous fire and the expense of the new shaft. This good ore came from an extension of an ore shoot directly on the dividing line between the two Homestakes which was being worked by the North Homestake through their 500-foot shaft already mentioned. We knew that they were mining ore close to the line, and that the fissure vein on which both ore shoots of the North Homestake occurred crossed the dividing line and extended into South Homestake holdings.

The north shaft, on the opposite side of the gulch

— John Kelt

BAXTER MTN. AND THE SOUTH HOMESTAKE

Main shaft is at top; ore-sorting building at bottom. The "blacksmith level" drift (tunnel) to the main shaft starts behind the lower building. The blacksmith shop (with smokestack) is at the right.

from the Devil's Kitchen, was started as an exploratory venture on the fissure vein at a point high on the dividing ridge and on the line between the two properties. Incidentally, the Devil's Kitchen deposit had no known connection with this fissure vein. The two are in different rock formations and of entirely different character.

No drifting was done until the 300-foot level was reached. Then we moved laterally toward the line. A drift 300 feet long was necessary, and it was hard going. Because of poor ventilation the last 200 feet were driven by a single shift. The work was slow and discouraging. We could hear the miners working in the North Homestake straight ahead and knew they were getting out good ore. Each day increased our apprehension that the shoot was all on their side of the line. When we were not more than twenty-five feet from the line and about ready to concede defeat, we encountered the ore shoot. With the first two rounds of shots we uncovered one of the richest pockets in the history of the district.

I was there when the shots were set off. We were all anxious to know the results and I was set up to begin panning anything we uncovered. Joe Grieshaber, our mine foreman, was standing on a track above the little platform where I had placed myself and my equipment. He was engaged with the other workmen and was not watching me, but I called him over as soon as I began working on the ore. He was a short, fat German with big, round eyes, and I thought those eyes would pop out of his head after one look into the pan. After another quick look, he grabbed the pan and doused it into the tub, quickly destroying all the evidence. "Don't say a word, Morris," he hissed, "not a word! You hang around or come back at quitting time and we will go down and see what the shots show. But not a word to anyone!"

There was no artificial ventilation in the shaft and it was not possible to work two shifts on account of the dust and powder smoke which followed the blasting. Several hours were required for the dust to settle and the air to clear. For that reason, only the day shift worked and the night crew did not come on at this shaft. Joe and I waited around that evening until the workmen had disappeared, going down to their work in the Number One shaft, and when we thought the smoke had cleared sufficiently, we went down the ladders, 300 feet, and back to the face of the drift. Joe carried an empty canvas sack over his shoulder to take a sample. The smoke was still pretty thick and we had difficulty seeing and breathing. There were no carbide or electric lamps in the mines in those days. All we had were the old-type candlesticks, sharp-pointed affairs about a foot long with a curved handle and a receptacle for holding the candle. The light was poor but we saw plenty.

The sight that met our eyes was a jewelry shop. Clean, bright, shining gold, big blotches, leaf gold and wire gold, coarse gold and fine, and a lot of it! With our candlesticks principally but sometimes using a pick, we sorted out the best, filled the sample sack (fifty pounds) and a powder box found on the level. These we carried back to the shaft; then climbed the ladders to the surface.

By this time it was away past midnight and we both had beautiful headaches. After getting ourselves calmed down somewhat, we lay down on the floor of the shaft house and slept. When the seven-o'clock crew came to work and the hoist was started, Joe had himself lowered to the 300-foot level and sent up the sack and box of specimens we had collected. He stayed with the men to gather what more he could while I took what we already had down to the assay office. The material was first ground in a hand mortar, then panned and melted. The amount we realized for our night's work was $18,000. That is what the company got. What the miners took for their share, nobody knows. We were satisfied — in fact, delighted.

From that time on the work progressed normally and well. We found that the ore shoot *raked south*. Above the 300-foot level the ore feathered to a point some twenty-five to thirty feet above the drift. Had the level been run at a depth of 250 feet in the shaft, a total blank would have been the result.

At the 600-foot level the shoot (300 feet long) was entirely on South Homestake ground. Only one other rich pocket was found. It yielded $3,000 as specimens. The average grade of ore mined from this shoot was approximately $8,000 per ton. Above the 500-foot level it produced $12 per ton; at the 800-foot level, $4 per ton. Below 800 feet the values were too low for profit. The average width of the vein was close to four feet.

For ten years the two Homestakes continued production, and the town prospered. The outlying districts were building up also with an increasing number of homesteads and ranches. Prospecting continued and new discoveries were made.

The time came when exhaustion of the ore bodies became apparent. Even then hopes that White Oaks would be a permanent town were reinforced by talk of a railroad, and the discovery of the Old Abe mine (of which more later) reinforced them still more positively.

Population of White Oaks, 1890, more than 2,000!

— *John Kelt*

UNDERGROUND — *a young girl, Dave Jackson, and Herman Kelt on the blacksmith level of the South Homestake mine 150 feet below the surface, with an ore car.*

As long as the mines produced, the town prospered, new businesses came into being, and there was healthy competition among the merchants. Occasionally the competition became almost too healthy. Trouble occurred, for instance, when a certain firm put in a stock of hay and grain. Paul Mayer, the livery-stable man, retaliated by purchasing flour in carload lots and discounting the price. The other side had to keep up, and soon flour was selling below the actual wholesale price — selling, as I remember, at ninety-five cents for a fifty-pound sack. It became the cheapest food commodity in town. Paul won the fight.

The stage was still our link with the outside world. E. W. (Father) operated the line until 1887-88, after which the stage line and equipment along with the mail contract — San Antonio, White Oaks, Lincoln — were taken over by Mr. Urbain Ozanne, owner and operator of the best hotel in town, the two-story brick Hotel Ozanne.

By 1890 the two sawmills had also changed hands, having been taken over by the Gumm brothers. They added a planing and shingle mill and a woodworking factory. This family (father, mother, five sons, one daughter) was one of the most interesting in town. Their family residence, built by themselves across the arroyo north of town, was the largest residence (two stories) and the most conspicuously located in White Oaks. The boys — Joe, Pete, Wall, John, Roy — were active in baseball, hunting, and other sports. Their sister Lovena (Vena) was popular in the church and social affairs of the community.

Noncommercial buildings included a second town hall, "Bonnell's Hall," used mostly as an athletic club for boxing, wrestling, general gymnastics, and for dancing, theatricals, musicals, and other local entertainments; and the two-story brick school building on the north side of the big arroyo. The school would have been a credit to a town much larger than ours. We had a kindergarten, elementary and grammar grades, a high school, and accredited college-entrance courses. Frank Richmond was the principal, and there were three or four other teachers. The attendance in 1890, exclusive of kindergarten, was sixty-five students.

Church affairs were an ever-present influence. By this time we had an adobe Baptist church led by Dr. and Mrs. A. G. Lane. These good people with their seven children, three boys and four girls (Will, Ida, Annie, Allen, Bruce, Mary, John), were among the town's most representative and popular citizens.

A friendly rivalry existed between the Baptists and the members of the Congregational church — a rivalry which aroused the energies of each and contributed to the interests of both. The first regular Congregational minister, a Mr. Hall, was sent out by the mission society in Boston. He was a small man, past middle age and somewhat lame. He was well educated, however, and sincere in his work. Unfortunately his sermons were far over the heads of his audience and lacking in interest for children. His wife was little help to him, either. A former schoolteacher, she was engaged to take charge of the White Oaks school (this was before the brick schoolhouse was built), but her health failed and she was unable to continue.

In later years the main difficulty was to get and keep a *good* minister, one who was a fair preacher. We learned that men of the clerical profession are human and sometimes weak. One, so they say, after "preaching himself out of old sermons," was lured to the hills by dreams of easy gold. He quit the pulpit and went prospecting, finally locating and settling on a claim in the Jicarillas ten miles north of town. He lived by himself in a one-room log cabin. One day he put several sticks of frozen dynamite in the oven of his stove to thaw, and then climbed up on the roof to fix the stove pipe. The explosion which occurred almost immediately threw him into the air and landed him on the side of a hill some forty or fifty feet from the cabin. Fortunately a clump of small pine trees afforded a cushion, probably saving his life. Thus he graduated from mining with "flying honors," came back to town and studied law.

There was still another, perhaps the most interesting speaker of all, who was likewise a victim of the mining bug. After some time in White Oaks, he endeavored to arrange for a crew to go down into Mexico on a prospecting trip. Suddenly and without notice he left for parts unknown. He took the stage to the railroad and then disappeared. His wife, an estimable and highly respected Englishwoman, managed somehow to reach Los Angeles with her son and daughter, ten and fourteen years old. She was able through perseverance and thrift to take care of and educate them and they both became prominent — the son an assistant superintendent of public schools, well known and well thought of.

After years of disappointments
Eureka, mining paid off.
Hard to get good clergmen,
after good gold fever.
Competition among businesses,
Churches

Days of Sunshine

IT SEEMS that my story of White Oaks cannot be written without references to the adventures of the Parker family and my own part in them, so here I will insert some bits of color and background from the common, everyday life of one family — details of the picture with which I was most familiar. I will begin by going back to the period which ended our three years in the log cabin and saw us moving to the eight-room adobe in the middle of the forty-acre tract on which grew the awesome leaning piñon tree used for a gallows.

The house was built by Will Patterson, one member of the ill-fated Little Mack-Henry Clay syndicate. It was well constructed, the walls eighteen inches thick, cool in summer and warm in winter, probably at that time the best and largest dwelling in town. Like all the rest, of course, it lacked plumbing, stationary sink, bath tub, and the other conveniences.

The balance of our household goods, stored in St. Louis, were sent for and hauled in heavy freight wagons

73

over the ninety miles of rough mountain and desert road
from San Antonio. They were a wonderful addition to
the meager furnishings of the log cabin.

The acres surrounding the house afforded good part-
time pasture for a cow or two and for the horses — well
bred and fast — on which Father's pride was always
centered. There was Polly, a big, fast trotting mare;
Prince, a picture stallion which under saddle would pace
and canter — a fine riding animal; the two gray ponies
Dick and Whitey, the best team and the best saddle
ponies in Lincoln County. Polly was later replaced by
Charlie, an easy-going but powerful black — a horse
which with reasonable whip encouragement could really
run.

Our favorites were Dick and Whitey. A hurry-up
trip to the railroad, ninety miles away, was once made
by Father behind these two in thirteen hours. Both were
superior saddle ponies also, a fact which caused many a
squabble among us three boys later on when we argued
about whose turn it was to use them for a buggy ride or
horseback frolic with a best girl.

The west half of the pasture, including a 200-foot-
high projection from the mesa to the south, was wooded
with oak, piñon, and cedar. Back of the house was a
pond supplied with water by the pipeline running from
White Oaks spring, two miles east, to the South Homestake mill, one mile west. An earth dam held back the
water which, at high level, covered about an acre. At one
side was a small ice house, a frame building with sawdust
filling. During the winter ice would freeze four to six
inches thick. With "whip" timber saws we cut the ice
into blocks about two feet square. We dragged it by
ropes to the ice house, stacked it inside, and covered it

with sawdust. We had plenty for our own use; often we had enough to share with others.

Mrs. McDonald often told the story of a young cowboy at the ranch who had never seen ice cream before. After consuming a dish of it, he remarked, "Gee, that's the coldest puddin I ever et."

We learned that if the ice house was not filled by Christmas, the chances were against getting any ice that season.

Back of the stable were three old Concord stagecoaches, relics of older days. Worn out, they were left there as junk. After 1890 very few, if any, Concords were replaced in the United States. Many were shipped or driven down into Mexico. As distances became shorter and the railroads absorbed transcontinental traffic, the coaches were replaced by lighter, less expensive, less comfortable vehicles. Steel springs took the place of leather rockers. Two-horse wagons displaced the four- and six-horse coaches. A very large part of the romance of long-distance Western travel disappeared with the Concord stages.

improved
Stagecoach

Our house had a wide covered porch across its full width, both front and rear. Close by in the yard was a dug well with overhanging pulley, rope, and bucket for drawing water. Beneath the floor of the back porch was a brick-lined, cement-plastered cistern for holding rain water from the imported shingle roof (the house was built before shingles were made locally). In time most of the better houses in town had shingle or galvanized-iron roofs and similar "soft-water" cisterns. Since well water contained alkali (lime, magnesia salts) it produced a heavy, sticky precipitate, impossible to avoid or remove, when combined with soap. As a result, cisterns were almost a necessity. Rain water was conserved for drinking, bathing, and the family wash. Water from the first rain of the season was usually considered to be roof wash to get rid of the wood, paint, fine sand, and dirt which would give a bad taste to the drinking water. Whatever went into the cistern was filtered through a half barrel filled with coarse gravel and charcoal. Water thus filtered was considered to be more palatable for drinking than water from the wells. All cisterns were equipped with an endless-chain pump with small iron buckets. When in use the inverted down buckets carried air into the agitated water and kept it pure and fresh.

With the larger house came the need for domestic help. When the boys were working or away at school, a maid was needed but almost impossible to find locally. Women of any age who wanted to "work out" were practically nonexistent, and therefore at a premium. We found and tried several. Their families were from Kansas, the Indian Territory, the plains of Texas. They were good cowgirls, perhaps, but ignorant of housework. Instead of helping Mother, they most likely added weight to her load.

One of our would-be helpers was a girl about seventeen years of age, one of a large family living on a ranch just outside town. She was a husky lass who could milk a cow, saddle a horse, chop wood, but her knowledge of housecleaning was nil and her cooking was limited to bacon, beans, flapjacks, and black coffee. Familiar only with tin cups and plates when she came to us, she was terrific on our crockery, china, and glassware.

Her background was showing on the first Saturday night she spent with us. In those days the kitchen was our bathroom and we bathed in one of the family's wash tubs, using water heated on the cookstove in a bucket or large teakettle. To start the new maid off right, Mother explained that the family took a bath each week and she could take hers before supper so the kitchen would be clear afterwards. The girl bathed as requested, then pushed the tub full of bathwater under the table. When Mother asked her why she did not throw out the water, she replied, "Oh, I thought you and Mr. Parker and the boys wanted to take a bath after supper." She was amazed when Mother told her to throw the water out, scrub and rinse the tub. And she had an argument. Ranch people had to economize on water, and her people, at home, always used the same bath water. The story soon spread around town with the caption: "The Parkers' Extravagance."

Our maid problem was not solved until 1887. This was two years after I had left home for school in Penn Yan, New York, Mother's old home. In 1886 Frank followed — to the manual training school at Washington University in St. Louis. In 1887 Mother came east for the summer, bringing Frank with her. At Penn Yan she found Louise Shanks, then about eighteen, one of two or three orphan children who were living with relatives. She was accustomed to housework and with a bit of training was qualified to meet our requirements. Since she had a good physique, was fairly attractive and was pleasant, she soon became an object of attraction. She lasted two or three years (longer than we had any right to expect) and finally married Professor Richmond, principal of the White Oaks school.

When Louise left, we were fortunate to get Christine, "Christy," Felnagel, a German girl who was the sister of our mine foreman at the South Homestake, Joe Grieshaber. She was a hard worker, loyal and dependable, an excellent housemaid and cook. She stayed until the South

Homestake closed down; then returned with Joe and his family to a farm in Kansas.

While all this was going on, we boys were away at school. During my three years away (two at Penn Yan and one in St. Louis), I studied Latin two years, Greek a year and a half, German one year, and French half a year. Spanish, the only language for which in later years I had any use, I studied not at all, neither then nor later; in White Oaks we had very few Mexicans. Languages and mathematics were my principal studies.

At high school in St. Louis the subject which interested me the most was chemistry. Perhaps because of my tales about White Oaks, Mr. Parsons, the instructor, dug out of the basement an old rusty gasoline furnace and endeavored to teach me the rudiments of fire assaying gold and silver. He was delighted to find that I already knew how to crush samples and make cupels. Trouble with the burner and lack of time before examinations at the end of the school year prevented me from achieving any real results, but I did find that this was a line of work in which I was interested.

When we left Grandmother and Grandfather Parker (with whom we stayed in St. Louis) in 1886 and returned to White Oaks, Frank worked in the mill as engineer and assistant mechanic. Fortunately for him he worked with Charlie Anderson, who had been our master mechanic for years. Charlie was a wizard at fixing things — anything of wood, iron or steel — without needing a supply depot or warehouse. He was a most valuable man for repair work, keeping the wheels going in spite of our distance from replacements. Frank did not return to school, but with his own industry and interest in machinery and with the training that Charlie gave him, he became known as one of the best practical mechanics in the country. Thanks to his experience in mill work and mining, his services were always in demand.

While Frank was serving his apprenticeship, I was beginning a career as an assayer. The assay office was vacant, so I took over the job. My salary was $50.00 a month, less than half the amount customarily paid for that work. I experimented for a week or so with the charcoal-coal furnaces and with formulae for fluxes found in the back of a price-list catalogue. I made cupels and practiced with the balances and metric weights; then made assays in duplicate to see that they checked. Soon the results of mine samples and mill tailings were again posted in the regular assay records. The work was later enlarged to include retorting of the mill amalgam and melting the bullion. Before the summer was over, the work was routine.

In society Frank and I were equally successful. We had been to dancing school in St. Louis. Our display of steps heretofore unseen on the floors of White Oaks placed us in demand, and since we were anything but backward, we paraded our accomplishments without stint. The ancient Virginia reel, the square dances, and the waltz gave place to the polka, the schottische, the York — with variations aplenty. With our city clothes and cultivated ballroom manners, we were a real sensation — or so we thought. Luckily there were enough girls to go around so we were not killed.

Some scenes from those days come back with special vividness. I remember seeing a young lady, a recent arrival, riding a burro astride, on a lot back of the house

MORRIS PARKER,
AGE 19

— *Lina Parker Mathews*

THE PARKER FAMILY
IN 1890

James, E. W., Morris, Mrs. Parker, Frank

in which she lived with her family. She was not yet an experienced rider and had some trouble guiding her mount. Ignoring her wishes, the burro walked unconcernedly under a taut clothesline tied between two posts exactly at the height of her chin. As chin met clothesline, the young lady clawed the air, fell heels over head, and made solid contact with the ground. Through tin cans, heaps of ashes, and other debris I sped to the rescue.

My experience was strangely similar to something that happened to Father. I have in the handwriting of his sister a note to the effect that Father first saw Mother getting out of a carriage. He took note of the prettiest foot and leg he had ever seen and proceeded to fall in love with them. "That's my wife if I can get her," he said. Evidence of keen eyesight and perfect judgment!

Father married his girl, and so did I, but it was not until some five years later. By this time my lady was a highly accomplished rider, perhaps the best in the vicinity, using a sidesaddle like the other ladies of the community. Riding astride was considered a vulgar breach of etiquette.

While Frank and I were away at school, Jamie held his grades under the tutelage of Professor Richmond. In the fall of 1888 the three of us went to Colorado Springs, where we attended college for two years. The first year we were there, Jamie bought a bicycle — a velocipede — one of these high-front-wheel affairs, a vicious thing to mount and ride. It cost $120.00, and was probably worth the money, for it certainly gave the rider a lot of hard punishment. The same bicycle could be seen, fifty-five years later, fastened to the wall of a store at Capitan (twenty-five miles east of White Oaks), with various other relics of early days in Lincoln County.

Because of my practice the summer before at home, I easily led the class in fire assaying. I was allowed to work with Mr. Lamb,[1] assistant professor of chemistry, in working out a correct fire assay of telluride ores, recently discovered at Cripple Creek. Assayers all over the country were having difficulty in getting results to agree on account of the volatile nature of the mineral. Mr. Lamb's experiments resulted in a simple formula since used on all tellurium ores.

On our return to White Oaks, Jamie got back his old job in the mill, and so did I. The assayer had quit to establish an independent office in town, so I took over — this time, as I recall, at $50.00 a month plus the income from any outside work that might come in.

Henry W. Kearsing, my predecessor, did not approve of the arrangement. He predicted openly that there "would be a mess with that kid in charge," and a little later on he attempted to prove his point. By this time, however, with two years of chemistry behind me plus hours spent in the laboratory and work with fire assaying, I was pretty well up on that sort of work.

Before many days had passed, a stranger brought in a paper sack containing a dark-colored finely pulverized substance. He wanted me to test it. The instant the hot flame touched it on a prepared piece of charcoal, it flamed a bit and disappeared. The smell of the fumes indicated organic vegetable matter. No residue remained, no coloring, nothing. I made a few simple tests. The

---

[1] Henry William Lamb was an assistant in the chemical laboratory at Colorado College from 1890 to 1894 (bulletins for 1887-1889 and 1895-1896 are missing). He was not an assistant professor in the strict sense.

results were nil. A favorite method of "salting" samples was to add a minute quantity of gold or silver in solution. To make sure nothing of this kind was going on, I made a careful assay. Not a trace!

When the man returned a day or two later, I handed him the certificate which read, "Vegetable matter. No Moisture." Down in one corner I had written: "Fee, $25.00."

His eyes flickered as he read but he quickly recovered and put on a poker face. "Say, what is this?" he said. "The only thing I can see is the fee — $25.00." He wanted to argue, but I held firm.

On the way home I stopped at Weed's store. Mr. Weed called me back to where he was sitting and wanted to know about that sample I had just assayed. I told him the story and he seemed greatly amused. When I mentioned the fee, his eyes bulged. Then he broke into a hearty laugh. Before I left he said repeatedly, "Don't you come down a penny. Make him pay." I wondered, naturally, what the joke was and what Mr. Weed knew that I didn't.

About a week later Mr. Weed handed me a check for $25.00 which I endorsed and he cashed. A few days afterward Mr. Kearsing left town. The story was out. Somewhere he had got hold of an old, dried-up potato, the kind prospectors carried in their pockets for months and years in the belief that it would prevent or help rheumatism. Supposedly, the drier it got, the more potent the effect. He ground it up to fine powder and had his friend bring it to me — for a laugh. Innocently the "kid assayer" provided the comic sequence. Henceforth in White Oaks my reputation as an assayer was *good*.

That summer was the beginning of a new life for

several of us. The Parker residence was a natural rendezvous for youthful entertainment, and we had birthday, anniversary, and holiday parties. Wisteria grew all along the front porch. There was a garden plot at one side with rustic seats and trees. On gala nights this garden was hung with garlands and lit by Chinese lanterns with candles burning inside. Moon or no moon, there were always bright stars above, and we had songs, games, music, and refreshments. Our house was a popular and congenial place for fellowship and pleasure.

In this environment Cupid loved to play hide and seek. Among those impaled by his arrows were my brother Frank, always in the vanguard, and Miss Bruce Lane, daughter of Dr. and Mrs. Lane. Genevieve and I felt the shaft and were married two years later, November 29, 1893, in the Congregational church by the Reverend E. H. Ashman of Albuquerque. Her father, T. B. McCourt, had passed away, and her mother was now married to W. C. McDonald of Carrizozo. During the days of our courtship Genevieve and I thought nothing of riding horseback or driving in a buggy the twelve miles between McDonald's ranch and White Oaks.

John J. McCourt, Genevieve's brother, followed our example a year later when he and Vena Gumm were married in the Congregational church. Our brother James, the slow and cautious one, held out for ten years longer, when, far afield in Manhattan, Kansas, he surrendered and was married to Miss Olive Sheldon, daughter of a jeweler at that place.

The greatest influence on our lives, and on the lives of many others during this time, was our mother. All through the seventeen years she lived at White Oaks she led the church and Sunday school activities and managed the various entertainments connected therewith. She was in church twice on Sundays — for morning and evening services — and was always present on Wednesday evenings at prayer meeting. Pleasant weather, rain, snow, dust, wind, or hot sun, she was there! As a member of the choir she attended choir practice, often twice a week, sometimes at home. She had a clear, well-modulated contralto voice and preferred to sing in that range, but often when it was necessary she sang soprano.

Several times a year children's entertainments were scheduled. The children were trained for their parts — speaking, acting, singing — under her supervision. Those entertainments are vivid in my memory. Comparisons I was able to make in later years with similar affairs in much larger towns showed superior preparation and production by the children of White Oaks. The people of our town, and of New Mexico in general, always gave merited credit to her untiring efforts in promoting the social cleanliness of the town, reflected, in later years, in the conduct of its young men and women. They still refer to her influence and cherish her memory.

I do not wish to imply that Mother was alone in this work or belittle the work of others. Actually all the outstanding women and their families did their part.

Along with Mother in the choir I remember Mel Hunter, a strong, deep bass; Frank Conger, tenor; and Vena Gumm McCourt, a capable and willing soprano, one of the most ardent workers in the church.

Among our favorite activities was the work we did with the White Oaks Dramatic Club. All we had was local talent locally developed — White Oaks kids grown up —

Three members in costume — James Parker, Wallace Gumm, Eugene Stewart

THE WHITE OAKS DRAMATIC CLUB

— Lina Parker Mathews

Playbill for **Among the Breakers**

but what a find they would have been for a present-day Hollywood scout! In our "treasure chest" we have photographs of characters in costume and we have the plays: *Nevada, or the Lost Mine; Among the Breakers; White Mountain Boy*. Printed descriptive programs are pasted on the backs of the photographs with the names of the characters. The men have mustaches and long, scraggly beards. The women wear skirts concealing their ankles (if any). And selected parts of the dialogue are reproduced: "YES, were you Mother Carey's old rooster, I'd marry you"; "Woman, fiend, you lie!"; "Oh, Peter, how can you before all these people!"

General admission was twenty-five cents, reserved seats fifty cents (the two front-row benches, exactly the same as all the others). Music by the town band, including Juan Reyes, a really good musician (specialty, violin and cornet); Al Green, (banjo); Billy Robinson, (flute); others whose names I do not recall.

Photographs were taken by M. H. Koch, the town's chief building contractor and carpenter, also its undertaker.

The club must have been successful, for it carried on for several years. Early-day mining camps (and campers) had a well-developed instinct for having fun. Home-made theatricals throve on the laughs of fellow citizens. The baseball diamond, with rival teams from different mines, was another, and noisier, high-time arena, but the Dramatic Club was our most civilized entertainment.

The Old Abe

THE OLD ABE was the biggest and best of them all! First located in November, 1879, shortly after the Homestake, it belonged to three Baxter Gulch placer miners, J. M. Allen, O. D. Kelley, and A. P. Livingston. They did their assessment work for three years. Failure to continue it in 1883 resulted in loss of title. The claim was open for relocation. In January, 1884, John Y. Hewitt, H. B. Fergusson, and William Watson located two claims: the White Oaks and the Robert E. Lee, covering in part the former Abraham Lincoln. Despite the new names the property has always been known as the Old Abe, and the name was retained when a company was organized seven years later.

The story is a good example of what happens to locators who ignore the legal requirements and teaches a useful lesson: work, or *somebody else will*.

Hewitt and Fergusson, attorneys, were both prominent in territorial and state affairs in after years. Fergusson especially distinguished himself, first as a long-time delegate in Congress, then as United States Senator. He was noted as one of the best speakers in the Senate. Mr. Watson, the third partner, studied law but was not particularly successful in practice.

The Old Abe is situated east and a trifle north of the North Homestake and of the fractional claim, the Lady Godiva, lying between the two. The wagon road up the gulch to the North Homestake passed directly across both the Old Abe and the Lady Godiva properties. For years, going to and from the North Homestake, miners and non-miners walked, rode and drove over the shallow covering of the ore body without realizing that a treasure was within scratching distance of their fingers.

The vein, a true fissure in felsite porphyry with no visible surface outcrop, offered little inducement to the owners, who halfheartedly continued to do the annual assessment more in hopes of a sale than in any expectancy of production by themselves. For eleven years after the first location the total amount of work done was just about equivalent to the assessment requirements for that length of time: 110 feet. There was a shallow drift seventy feet long with a shaft (winze) about forty feet in depth near its face. This work was on a gentle slope within 150 feet of the road.

During the fall of 1890 the owners offered a contract-option for purchase of the property to the South Homestake. Price: $40,000; work to start within thirty days; $10,000 to be paid six months later; the balance twelve to twenty-four months thereafter. Suspicion was strong that the main reason for such easy terms was a desire to get the current assessment done without cost to the owners.

Samples were taken by Joe Grieshaber and assayed

THE RICHEST GOLD MINE — *The Old Abe*

by me. The results were more than satisfactory — double
what we expected. A representative of the South Homestake made a hurried trip to the railroad and took the
contract to St. Louis, where it was rejected, thumbs down.

When it was definitely decided that the South Homestake would do nothing with the option-contract, the
partners had to do something about the annual assessment. Mr. Fergusson was living in Albuquerque and could
do nothing himself, but Watson was there. He was a
good-natured fellow though not much of a lawyer, apparently always hard up for money. He had with him a
nephew named for him, Watson (Watt) Hoyle, a recent
arrival in White Oaks. That fall these two men, Watson
and Hoyle, worked with pick and shovel on the county
road for three days rather than pay the three-dollar poll
tax required for voting. Their next job was the assessment
work on the Old Abe. Fergusson and Hewitt put up two-
thirds of the hundred dollars in cash. Watson was to do
the work, giving Hoyle half the money and one-fourth of
his one-third interest in the mine — one-twelfth of the
whole.

The easiest and cheapest place to do the work was at a point just below the dump from the previous work, in line with the tunnel and about seventy-five feet from the road. Here they sank a ten-foot hole. Encouraging values induced the owners to continue the work. When the hole was thirty feet deep, they had a pile (dump) on the surface, which I sampled and assayed. The tests showed an average of $27.00 per ton.

With a hand windlass they continued to a depth of seventy-five feet, the vein increasing in both value and width. Arrangements were then made to lease the old Glass ten-stamp mill and put it in working order.

— John Kelt

THE OLD ABE MILL, INTERIOR VIEW
*Bill Kennedy, second from left; D. L. Jackson,
center; Watson Hoyle, second from right.
The other two men are unidentified.*

About this time Rolla Wells, president and principal owner of the South Homestake, visited White Oaks. The property was still available, and he knew what was taking place, but again he refused to have anything to do with the Old Abe venture.

The mill proved successful, the recovery surpassing anything yet known in White Oaks. As profits came in, a hoist was installed, the shaft was enlarged and timbered, drifts were run, and the mine went into steady production.

The assessment work which opened up the Old Abe was done about the first of November, 1890. During May of 1891 Mr. Wells was again in White Oaks and is said to have offered $300,000 spot cash for the Old Abe property. The owners smilingly refused and doubled the figure. Why sell? They were on their way and satisfied to hold on.

The North Homestake was then in a period of short production and was willing to lease the mill for use by the Old Abe. The two mills treated about fifty tons of ore daily, and *net* profits were close to $1,000 per day. For a period of two months they were more than double that figure. The *tailings* discharged approximately $18.00 per ton, twice the value of the *ores* mined by either of the Homestakes.

Tailings from the Glass mill flowed down the creek and were lost. Those from the North Homestake were impounded and a few years later were treated by the cyanide process. The returns were a welcome addition to our own $2.00- and $3.00-grade tailings. Included was one lot, 1,500 tons, of mixed North Homestake-Old Abe tailings which averaged $10 per ton, recovered value.

For fuel the Old Abe opened a coal mine of their own about half a mile from the Parker mine, presumably on the same vein. Thus they had an independent supply of excellent coal at very low cost.

About six months after the Old Abe started production, on July 1, 1891, the terrible mine fire occurred at the South Homestake wherein two miners lost their lives, the hoist machinery was destroyed, and the shaft timbers were totally ruined. Coming so soon after the Old Abe mistake, the fire caused the St. Louis stockholders to lose interest in their enterprises at both White Oaks and Nogal. It hurt them to realize that $1,000 spent in the forty-foot shaft-winze at the beginning of the work on the Old Abe would have produced sufficient high-grade ore to pay the purchase price plus all operating costs without the expenditure of another dollar. No wonder they were sick! Such is mining — sometimes!

The story has a brighter side in that local men, no outsiders (unless Mr. Fergusson, a former resident, might be so classed), reaped the full benefit of the Old Abe, the biggest and best mine in the district.

The end of operations at the South Homestake was averted by leasing the mill to the Old Abe. Shortly after the deal was completed, it became the only mill in operation. The Glass and North Homestake mills were closed down and the company worked on better amalgamation, continuing in this fashion for a year or more before building a mill of its own. In the meantime, using borrowed money and profits from the mill-lease contract, the South Homestake was able to sink the new Devil's Kitchen shaft and run drifts to connect with the lower levels of the burned shaft, necessarily abandoned on account of the fire.

The discovery of the Old Abe meant a great deal to the life of the town and to the continuation of mining at White Oaks. The two Homestakes were now at critical depths, short of ore and apprehensive of the future. Uncertainty was now changed to assurance and hopes rose for production beyond that of any previous period. Mr. Hewitt, an able attorney of high integrity, loyal to his principles, assumed the management of the Old Abe and the leadership of town affairs in general. Mr. Watson, who had married the daughter of the "dynamite" preacher, continued in his role of general utility man. His nephew, Watt Hoyle, went through a training period as amalgamator and was placed in charge of the mill.

Employment was fairly distributed, new faces appeared on the streets, and the town took to itself a degree of prosperity theretofore unknown. It was during this period that the population reached its peak — approximately 2,500 persons.

New dwellings were built and many improvements were made. A bank was organized and a two-story building was constructed for its accommodation. As I recall, the bank was a branch of the San Miguel County Bank of Las Vegas. Joshua Raynolds[1] was president of that bank, and of the White Oaks bank. Vice president was Mr. Hewitt; Frank Sager (his nephew or cousin) was secretary and treasurer; George Ulrich was one of the directors. Later Mr. Hewitt assumed the presidency with Mr. Ulrich as vice-president, Mr. Sager as secretary and treasurer. Long after the bank was moved to Carrizozo (in 1907), they retained their positions.

Incidentally, when the bank building was planned, several wagonloads of cut stone were hauled from the South Fork of the Bonito River some thirty miles distant. The colorful banded stone was supposed to be marble, but on examination it proved to be alunite, a hydrous sulphate of aluminum and potassium, wholly unsuited for

---

[1] Joshua Raynolds (1845-1932) originated in Ohio but became a banking leader in Colorado and New Mexico. He lived at Las Vegas but had banking interests in Albuquerque and El Paso.

— *George W. Shalkhauser*

WHITE OAKS LEGEND — *The Hoyle House*

building purposes. Fortunately the error was discovered in time, but the episode brought chagrin to the bankers and a laugh to the townspeople.

The new prosperity was not good for everybody. Watt Hoyle, for instance, fell into temptation as a result of his one-twelfth share in the Old Abe. He decided to build a "superior" residence — a two-story brick, the finest in town. A bachelor, he lived there with his older brother Will and Will's wife as long as the mine continued to produce. Yet the interior of the house was never finished and it was never completely equipped and furnished. Before long it was dubbed "Hoyle's folly." [2]

Several less pretentious brick and adobe structures were built and others were enlarged. New or improved buildings included Jones and Taliaferro's store, Paul Mayer's livery and feed stable, Ziegler's, Solomon's, Ridgeway's, and Stewart's stores, Mr. Hewitt's adobe residence, Mr. Sager's house (a particularly choice one), Mr. Watson's two-story adobe, and a score of others — a credit to any town, anywhere.

At the height of this complacency and prosperity catastrophe struck the town for the second time — the greatest tragedy in its history. This was another mine fire, in the Old Abe, in which nine miners lost their lives. It occurred just two years after the fire at the South Homestake.

The Old Abe fire started from the explosion of a kerosene lamp hanging on the wooden frame of the hoist house. Almost immediately the place was a mass of flames, quickly mounting the gallows frame over the shaft.

---

[2] The Hoyle house was completed in December, 1892 (John Kelt to C.L.S., February 27, 1969).

Twenty men were working underground at the time, eleven of whom succeeded in climbing the ladders to the surface. The drift at the 300-foot level extended north from the shaft about 400 feet with an upraise and ladders to the surface at the far end. To reach this escape level from below, it was necessary to climb the ladders of the shaft.

The lowest level, as I recall, was 800 feet, where several men were at work not far from the shaft. Others were in the stopes on the 700-foot and 500-foot levels, 100 to 150 feet from the shaft and thirty to fifty feet above the track. Failure of the cage to return to the 700-foot level, from which ore was being hoisted, caused the station tender to pull the signal cord. He discovered that the cord was broken. Looking up the shaft, he saw the flames raging in the gallows frame. First calling to the men below and getting their reply, he rushed back to the stope on the 700-foot level, shouting the alarm; then he dashed back and up the ladder to the 500, where again he ran back to warn the miners, and so on to the 300-foot level. Up the ladder one by one the men struggled.

Climbing a ladder 500 feet in a vertical shaft is hard, breathtaking work for any strong man, even under normal conditions with plenty of time for rest at each 100-foot station. It takes a brave spirit and a fighting heart to keep on going when one is short of breath in air fast being depleted of oxygen. The big wonder is that so many of the entrapped twenty men were able to escape.

Later on several bodies were recovered in the ladderway of the shaft. Some had fallen back to the station landings. One was hanging by an arm locked back of the ladder against the timbers. Others were found along the drift at the 300-foot level, face down with arms out-stretched toward the exit raise. All showed evidence of having been overcome by monoxide gas.

In addition to the loose supply on hand, air pipes were torn from installations of the North and South Homestakes and the Lady Godiva. These were strung along the surface, then down the exit, and back to a connection with the mine shaft. All the compressors in camp were pressed into service to aid in the air effort.

Water finally overcame the fire, and air clarified the workings, but it took two days and three nights to do the job as the townspeople, men and women, fought against terrible odds. They worked in relays battling flames, smoke, and gases as they attempted to rescue lost relatives and friends or recover their bodies. In the meantime tents were erected and women from town, working in shifts, continuously served coffee and meals. Selected groups of men worked without stopping until they returned from the front exhausted to catch an hour's rest on the cots and bedding provided for them. They refused to stop long enough to regain their strength. All were friends or relatives of the unfortunate victims.

The funeral was a particularly sad event. Practically the whole town was there, overflowing the church, which was far too small for the attendance. Since there were not enough factory-made coffins, the undertaker engaged a couple of carpenters and they supplied the deficiency with pine boards planed, fitted together, covered outside with black muslin and inside with white. Nine coffins, side by side, filled the width of the church. After the music and the service the bodies of those young men — active and full of life a few days before; now blackened, scorched, some of them unrecognizable — were carried out one by one and placed in open lumber wagons, two to each. Then the funeral cortege, including every passenger vehicle in town — buggies, hacks, light spring wagons — strung out along the dry, dusty road leading to the cemetery west of town. It was more than a mile west of the church, but many people followed the vehicles on foot.

The graves were ready, mounds of dirt alongside. At each there was a short service and a prayer. Then the coffins were lowered, tops were placed on the rough pine boxes enclosing them and screwed down. Then the comrades and fellow workers of those departed filled the graves, placed the wooden markers, and shaped the mounds.

Not all the men were single, and their families, burdened with grief, silent and helpless, overcome by the tragedy, were a heartrending sight.

Weeks of grief and readjustment followed. The whole town suffered. Naturally all mothers and wives were reluctant for their loved ones to face again the hazards of underground work to which they had formerly been accustomed.

Realizing that the conduct of the mine and mill now required more knowledge and experience than they themselves possessed, particularly since they were negotiating for the construction of a mill, the owners employed J. B. Risque as manager. He brought with him two assistants: Benjamin B. Thayer [3] as mine foreman, and John A. Foley as master mechanic. All three were fine, cultured young men. Risque and Thayer were graduate mining engineers from Columbia University and Foley, as I recall, was a

---

[3] Benjamin B. Thayer (1862-1933) was a noted mining engineer who was president of Anaconda from 1908 to 1915.

graduate in mechanical engineering of the Massachusetts Institute of Technology.

Mr. Thayer rented the two-story adobe Heman residence near the southeast edge of town and moved in with his wife, her sister and her mother. A few months after their arrival, Risque married the sister. This was in November, 1894.

They went to work at once. The Old Abe mill, a near-duplicate of the South Homestake mill, was built near the mine. A bored well in the gulch below the North Homestake mill supplied water. It was pumped over a mile and elevated more than five hundred feet. The vertical, two-compartment mine shaft was enlarged and equipped with a larger hoist, guides, and a cage — the only cage in the district. The mill was entirely successful. Later, a cyanide plant was added to treat the tailings.

Shortly after the mill began operations, possibly because a labor shortage seemed imminent, the new management decided to raise wages from fifty cents to a dollar a day. To both the Homestakes, this was disturbing news, since neither was in a position to follow suit. Naturally the best workmen in camp were attracted by this increase. The difficulties of the Homestakes, already at near peak, were increased, as were their expenses. Many people believed that the situation contributed to their closing down at an earlier date than would have otherwise been necessary.

It was, as I recall, about eighteen months after the new management assumed control that Mr. Thayer, the mine foreman, was let out — *fired* — by Mr. Hewitt. He was blamed for extracting ore from the vein too close to the shaft, thus causing the timbers to collapse and the walls to cave, destroying the alignment at that point.

It was the best thing that could have happened to Mr. Thayer. On leaving White Oaks, he went to Silver City, New Mexico, where he got a job with the American Smelting and Refining Company at their Pinos Altos plant. In time he became superintendent. For the balance of his life he remained associated with the affairs of that company, eventually advancing to the presidency with offices in New York City. He became one of the most prominent men in his profession — head of one of the largest and best-known mining organizations in the world. The Old Abe was a steppingstone on the trail which led Ben to the summit.

When, some time later, Mr. Risque also left White Oaks, he worked under Mr. Thayer, their former positions reversed. Afterward he opened an independent engineering office at Salt Lake City, where he too became highly successful.

To his credit, Mr. Hewitt, small-town lawyer and mine owner, added a notable footnote to mining history when he — justifiably — discharged the future top man of them all.

Mr. Foley stayed on at White Oaks for four or five years, the last two with the South Homestake. He married Mary, youngest daughter of Dr. and Mrs. Lane. Thus there developed a warm friendship with my brother Frank who, in 1882, had married Mary's older sister Bruce. When Mr. Foley left White Oaks, called by a San Francisco mining man to dewater a deep, old-time gold mine at Quartzburg, Idaho, he sent for Frank to assist with the work. Frank, with his family, remained in the Quartzburg-Boise district five or six years.

The Cyanide Process

1893- First Cyanide Plant in New Mex

In 1892 the McArthur-Forrest Process, more familiarly known as the Cyanide Process, was brought into the United States from Australia. A testing plant was built near the outskirts of Denver.

The process involved the principle that gold is soluble in a weak solution of potassium cyanide, from which it can be recovered by precipitation on metallic zinc. After successful operation in Australia, it was patented in other countries of the world, including the United States. Later, after being contested by United States mining companies, the patents were annulled.

Application of the process at that time was confined to percolation, commonly referred to as *leaching*. Treatment was limited to the sand product of mill tailings. The light mud (slime) was discarded because there was no known means of filtering to clarify the solution. In later years, however, mechanical filter presses were designed, after which the all-slime treatment was preferred. The sands were reground to fine powder. The process was also

102

applied to silver ores, native and chloride. Sodium cyanide, as efficient as potassium cyanide and less expensive, was later on used almost exclusively.

In June of 1892, when I was on my way home from college at Rolla, Missouri, I stopped off at Denver to wait for my brother James, then a student at the Colorado School of Mines at Golden. While in Denver I visited the McArthur plant and became convinced of the merit of the cyanide process in the treatment of gold tailings. After reaching home, I arranged to have samples of South Homestake tailings sent to Denver and tested. The results were highly satisfactory.

The South Homestake tailing pond, an acre or more in extent, was measured and carefully sampled, and found to contain about 20,000 tons of average gold value — two dollars per ton. A one-ton sample was shipped to Denver by freight. I followed to watch the tests, and thus became more familiar with the details of the operation. I returned to White Oaks determined to build a plant to treat South Homestake tailings. The next year, 1893, I built and operated the first cyanide plant in New Mexico, one of the first in America, at White Oaks.

It consisted of six wooden tanks, each one twenty-two feet in diameter with four-foot staves. For each there was a solution and sump tank sixteen by six feet, a precipitation box for zinc shavings, a small rotary pump, and a gasoline engine. The solution used contained five pounds of potassium cyanide to one ton of water (one-fourth of one percent). Consumption of cyanide was about one pound per ton of tailings treated. Extraction from clean sands (no slimes) was approximately ninety percent. The capacity of each leaching tank was thirty tons; cycle for leaching, five days; for unloading and filling, one day. A breakdown showed that operating costs were $.60 per ton; recovery, $2.60; profit, $2.00.

In 1894 a similar plant was completed for the North Homestake, thus giving me credit for the first two in New Mexico. We had no contract with the McArthur-Forrest people; hence they collected no royalty.

The North Homestake pond contained about two-thirds the tonnage of the South Homestake, but since primary settling was done in V-shaped iron cars, there was a higher percentage of clean sands. From one batch (mixed Old Abe tailings) we recovered an average of nearly $10.00 per ton, highest of all White Oaks tailings treated by cyanidation.[1]

It was due to the success of these two plants that the Old Abe built a similar one below their mill.

---

[1] The White Oaks *Eagle* for February 5, 1903, notes that "Thirty-two thousand five hundred and fifty tons of tailings from the South Homestake mill have been cyanided up to the present and out of this 32,550 tons, $81,375 in gold, has been saved by the cyanide methods used."

[handwritten: extracted]
[handwritten: Cyanide method salvaged that gold which would other wise have been wasted]

Bright Yellow Gold

1893 — WORLD'S FAIR YEAR in Chicago! A fairly representative collection of mineral specimens from Lincoln County, New Mexico, was placed on exhibition: gold, coal, iron, lead, zinc, copper, and gypsum (selenite) with native sulphur. The first three received honorable mention, the gold specimens being among the best there. I was in charge of the exhibit.

I remember one particularly fine specimen from the Old Abe — a mass of gold wire in a large cavity in a rock about the size of a coconut. The intrinsic value was probably not more than $100, but I was offered $1,000 for it and wired the offer to Mr. Hewitt. He declined on the grounds that the Old Abe had a standing offer of $300 for the arrest of anyone found in possession of a specimen taken from the mine. In due time this particular specimen was returned, put into the mill battery, and crushed with common ore. I remember that at the time it seemed like sacrilege, an unwarranted destruction of nature's art. I am sure I could easily have shed tears if I had seen

105

that beautiful specimen disappear beneath the pounding stamps.

This brings up the subject of specimen gold and what happens to it. Samples taken from the richest and most interesting gold-bearing formations are, of course, valuable, and they have a magnetism that few can resist. Actually, however, they are sought not so much because of their dollar value but because they are personal prizes — something to display. The pride of ownership is paramount in the minds of everyone involved, from the miner who sees the specimen first and pries it from the solid rock to the capitalist at his mahogany desk who likes to have a free-gold specimen in the upper left-hand drawer. The total value of specimens thus held by prideful owners would, if converted into bullion, amount to millions of dollars.

To the mine owner, of course, specimen ore is of value only because it helps to sustain the average grade of the ore mined. But he too is proud of his specimens, likes to show them off, and often gives them away as presents from one good fellow to another, the recipient unmindful of the dollar value of his present.

The bulk of the world's gold, as every mining man knows, does not come from high-grade specimen mines. It comes from those of large tonnage and low gold content — from ores which contain less than half an ounce of gold per ton. The proportion can be as low as one to 58332. Of late years a very large proportion of the gold produced has been a by-product of the smelting of copper, lead, and silver ores.

The White Oaks mines produced many magnificent specimens from the felsite deposits — bright yellow untarnished "leaf" and "wire" gold, some of it crystalline,

100 percent pure; coarse, heavy gold in the tight seams of the South Homestake diorite, some of the specimens weighing out more than fifty percent gold. These latter specimens were often associated with black crystals of huebnerite, a manganese-iron-tungsten mineral.

The presence of these rich specimens offered an almost irresistible temptation to unscrupulous workmen. All specimen mines everywhere are subject to this sort of "expropriation." The value of the gold filched from the workings at White Oaks was unquestionably great, and it went on in spite of all precautions and the many traps devised to prevent it.

The miner's conscience was not particularly involved. In his view he had a right to whatever "stuck to his fingers." If he got caught, he might be discharged, but that was all that was likely to happen to him. The law imposed severe penalties, but if the man were to be convicted, there had to be corroborating evidence — *proof* that he was seen to have taken the identical "Exhibit No. 1" from the particular property claiming ownership. This was a condition almost impossible to fulfill.

Actually at White Oaks we never saw evidence of retail or wholesale marketing of specimens the way it happened at Cripple Creek and other places, though open display of special prizes among friends was not uncommon. And every now and then we had reminders of what was going on.

A miner living by himself in a one-room cabin died of pneumonia. He had not lived in White Oaks long and had been employed at only one of the mines. When his body was removed, the removers found a box under his cot which contained a number of gold specimens. Because there were no other claimants, the authorities released them to the company which had employed him. Melted into bullion, they turned out to be worth approximately $700.

Of course not all miners were of this type. I personally know of cases in which a miner brought an exceptionally fine specimen to the office with a request that he be allowed to keep it. The request was granted without any stated or implied reservations.

When the mill work involved handling amalgam, a process in which the volume was less and the value was greater than with specimen ores, the owner was in a bad fix if he had a dishonest mill man. The amalgamator-battery man obviously had to be one who deserved complete confidence and trust since he was responsible not only for proper mill operation but also for delivery of the gold recovered.

During the twenty years or more of White Oaks activity the only known incident involving abuse of trust in this position occurred during the early years of operation at the South Homestake mill. For some three and a half years the head amalgamator worked a twelve-hour shift at $3.50 per day — top wages. Eventually he left New Mexico and moved to California, where he became the owner of a Los Angeles orange grove (worth $8,000) and bought a half interest in a saloon and wholesale liquor store (valued at $10,000). A detective agency was engaged to ascertain, if possible, the source of the money involved in these purchases. The expenditure of more than a thousand dollars and several months' time brought no significant clues and the matter was dropped. But our suspicions lingered on.

The only outright attempt at bullion theft occurred about 1890. Again the South Homestake was involved.

A routine cleanup at the mill produced a gold brick about half the size of an ordinary building brick. It weighed about twenty-one pounds and was worth $5,000. It was taken downtown and placed in the heavy steel vault in Mr. Weed's store. It was to be shipped out the next day. During the night the store and safe were broken into and the brick was stolen.

There was much excitement for several days but no clues. There were no strangers in town, nobody to be suspicious of. All that could be done was to detail three or four men to keep careful watch around town, concentrating especially on persons going out and coming in.

This policy bore fruit some days later. Two of the watchers (the night crew) heard strange noises coming from the back room of a hardware store on the main street. The noises sounded like somebody pounding on steel. The watchmen waited to see what would happen.

Shortly before daylight two men came out of the back door, harnessed a team to an old buggy, and drove off, the two watchers following on horseback. Three miles out of town, just after daylight, the buggy left the road and turned into a cluster of trees. The men got out, dug a hole, and buried something — then returned to town.

The buried article turned out to be an old iron casting. Examined, it showed fresh chisel marks and cuts. In the latter, small fragments of metal, apparently gold, could be seen. These fragments, some of them no larger than a pinhead, were collected and I assayed them. They proved to be of the same fineness as the missing brick, which I had previously sampled and assayed.

That same afternoon Mr. Langston, the town constable, came to the store with four deputies. Entering at both ends, they arrested the owners, who proved to be the two men with the buggy. They at first denied everything, but a search of the workshop and the shelves soon turned up several tin cans and other receptacles hidden here and there which contained portions of the chopped-up brick. Seeing that it was no use holding out in the face of this evidence, the men confessed and got a term in the penitentiary.

The $5,000 brick, incidentally, was intended to be sent to the railroad by special conveyance, probably in care of some trusted friend of the mine manager. This was a fairly common procedure and was the result in large measure of a limitation on insurance imposed by the stage company. Bullion shipped by stage could be insured only to the amount of $1,000, and a premium or fee of $10 was charged. At the South Homestake it was customary to melt the product into four-pound bricks valued at approximately $1,000 each. The bricks were first wrapped in paper, then in heavy white canvas. They were tied, wax sealed, and addressed to a Denver bank. The bank delivered the brick to the United States Assay Office for assay and the net value was placed to the owner's credit in the bank.

For years "dummies" of brass, exact replicas of the gold bricks, were carried by the stage in case of holdups, but they really served no useful purpose. During the period of White Oaks operations the daily (each way) stage was held up only once. The robbery took place near the end of mining activity when shipments were irregular, and the robbers happened to pick a day when the stage carried no bullion.

The perfect record of the White Oaks mines with regard to holdups was partly a matter of luck, but the shippers tried hard to take proper precautions. One special example that I remember happened on August 20, 1891. The schedule called for the stage to leave White Oaks at three P.M. and arrive at Carthage before noon the next day — ninety miles to the railroad; twenty hours on the road. I was on it, leaving for my last year of college, but it was a special trip in other ways. For one thing we had had a wedding in White Oaks, and Mr. Watson and his bride were getting ready to leave for a honeymoon outside. They were not on the stage themselves, but their trunks were — four of them — with the usual lot of express and mail sacks, a heavy load.

Just a few days before, the Old Abe had finished a cleanup representing two weeks' run and the treatment of five hundred tons of ore — the first since they had leased the South Homestake mill. The product was sixty-five pounds of gold worth over $16,000.[1] Just before the stage left, I learned that the Watson trunks contained the recent cleanup, distributed equally among them.

The trip started badly. Shortly after we left town, we ran into a downpour which left the road deep in mud. Four horses pulled the coach, but travel was slow and tiresome.

Well past midnight we reached the Oscura range. The road was winding and hilly and broken by dozens of *arroyos* with steep banks which called forth double-strength profanity from the drivers and every ounce of strength and energy from the horses. Finally we reached a slope where the tired animals could go no farther. The stage stopped.

---

[1] The Socorro *Bullion* reported in the issue of November 1, 1883: "On the 11th of last month the White Oaks stage came into San Antonio with $22,000 gold bullion."

I got out and walked to the top of the hill, possibly fifty yards. The sky was clear; the moon three-quarters full and approaching the western horizon. It cast shadows among the scattered trees — shadows which to an imaginative mind might mean anything. Here, I thought, was an ideal place for a holdup — forty miles from nowhere. The only habitations were the stage stations for changing horses, twenty to thirty miles apart and operated by a single man, or at most two men. Ranch houses were few and remote. A robber could have everything his own way.

Nothing happened. The stage was backed to the foot of the hill. The horses were allowed a few minutes to rest. Then up the grade they came. It was as simple as that. Even so, with the moon where it was, going down, down, down, every fair-sized cactus along the road looked like a man; every big rock and every tree concealed a bandit ready to jump out and yell, "Hands up!"

At Carthage the station agent, who had received the keys the day before, opened the trunks and from the assortment of honeymoon garments retrieved the sealed packages (retorted amalgam, not melted into bricks), weighed them on an ordinary platform scale, and put them in the regular Wells Fargo safety chest for shipment.

That was the largest single shipment of gold I ever saw from the White Oaks district.

Nobody was killed, nobody was robbed as the gold went out, but characters who would have been at home in Western literature did occasionally break our otherwise clean record. There was the time when Bill Hudgins, a White Oaks saloonkeeper, shot and killed Jim Redman, a harmless young cowboy and man-about-town. Hudgins followed him outside the saloon to do the killing, but at the trial he was released on a verdict of self defense.

There was also old Chew, the laundryman — the only Chinaman in town — who was stabbed while sleeping. He was supposed to have money hidden. Suspicion pointed to a well-known young man. Blood was found on his shirt and overalls, but this was "circumstantial" evidence, not sufficient for conviction.

Episodes like these were few and far between, and life in White Oaks was pretty much routine. At the South Homestake my job, in addition to the mill work, was assaying — retorting amalgam and melting bullion. The monotony of retorting, cutting the retort gold into four-pound units, melting the bricks, pouring them, wrapping them, marking them for shipment — these are among my most tiresome memories.

The gold at times was irresistable & the miners. Pilverage.
Robberies

Mine fires tragic
Two men killed i one
Nine men killed in another

Fadeout

As EARLY AS 1893 the owners of both Homestakes were thinking about selling out and the mines were temporarily consolidated just before the Old Abe fire for the purpose of making a combined sale. Various brokerage houses and prospective buyers were notified. Joshua Raynolds, the Las Vegas banker, was granted the first option and through him a Mr. Martime made an investigation with a view of presenting the properties to certain of his clients.

When these clients heard that neither property could produce a survey or an assay map and that the owners were unwilling to go to the expense of having either one made, they lost interest. Three or four others — J. W. Yocum, and Yankee and Johns among them — refused to negotiate for the same reason.

As a result I was instructed by Father and Mr. Lloyd to go ahead and make the surveys and maps. I had two helpers underground and the hoist engineer at the surface and we started in. The North Homestake alone, in which most of the available ore was exposed, required about six weeks to do the assaying and write the report. It took a full two months to complete all the work.

Now that this collection of data was available, Haggin and Hearst,[1] big-time operators of that period, became interested and they sent Frank Keller of the New York engineering firm to investigate. He spent several weeks in White Oaks. He checked the maps and reports. And he *recommended the purchase*. Because of outside interference, however, the deal was not consummated.

The end was not far off for the Homestakes. When the Old Abe fire occurred, the South Homestake workings were about ready to resume production. The mill, now released by the Old Abe, was put to work on South Homestake ores, mainly from the new shaft on the Devil's Kitchen ore body and the drifts connecting with the lower levels of the old burned-out shaft.

The activity was only temporary. After a few months the available tonnage was insufficient to keep the mill busy at its twenty-four-hour capacity. The mill operation became intermittent and was limited to twelve-hour shifts daily.

Other factors besides the ore shortage seriously affected the ability of both Homestakes to operate at a profit. They included increased depth, lower grade of ore, and inadequate surface equipment, especially hoists and compressors.

The North Homestake closed first, final operations

[1] Senator George Hearst (1820-1891) and James Ben Ali Haggin (1827-1914) were partners in various cattle and mining enterprises.

being confined to cyanidation of the tailings. The South
Homestake struggled a couple of years longer. Then, having disposed of their properties at Nogal (such as the
Helen Rae), the owners of the South Homestake sold
the mine and the coal mines east of town to three young
men resident in White Oaks: Allen Lane (mechanic),
Ed Queen (miner), and David L. Jackson (mill man).[2]
The price was low and the terms were easy — no doubt
to assure the sale.

Employing a minimum of labor and doing most of
the work themselves, the three young men were successful — so much so that they made a similar deal with the
North Homestake. A short crosscut in the latter disclosed
a diagonal vein branching into the diorite (supposedly
the Lady Godiva vein) from which they extracted a considerable tonnage of good ore. A couple of years after
the partnership was formed, Ed Queen dropped out and
left town. Lane and Jackson continued with varying
success.

When the Old Abe finally shut down, shortly after
the turn of the century, Lane and Jackson were instru-
mental in consolidating the coal mines and organizing the
utility company which for years (until 1942) supplied
electricity to the town of Carrizozo, twelve miles west.

Watson Hoyle left White Oaks and went, as I recall,
to Denver, where he lost most, if not all, of his money
to real estate and other sharks. Mr. Hewitt remained and
spent most of his in a futile attempt to revive the already
worked-out ore body. His last few years he passed quietly
at White Oaks among his old-timer friends and acquaint-

---

[2] The company was called the Wild Cat Leasing Company.
George Queen was a partner but dropped out after two years
(Mrs. C. L. Wetzel to C.L.S., January 4, 1970).

ances in the environment he loved, and there he passed
away, well past ninety years.

Dr. Paden, the town's standby as physician and
surgeon, moved to Carrizozo and continued his work as
general practitioner and as chief of the railroad hospital.
He was always well liked and respected, but he made
sure of keeping abreast of his profession by taking
brush-up courses at Rush and Johns Hopkins. When
finally old age compelled him to slow down, he too
returned to his old home in White Oaks. Well past ninety,
he was laid to rest there, long after his wife and two sons
had passed on.

With the closing of the mines and the death or
departure of so many of the townspeople, White Oaks
faded away in the early 1900's. Its last hope was the
railroad, which had to pass through White Oaks Gap
(the wise ones said) and would some day bring permanent prosperity to the town. Had their dreams come
true, White Oaks would not be what it is today.

The eighties, nineties, and the first decade of the
twentieth century were the great days of railroad building west of the Mississippi River as the transcontinental
lines added secondary branches to their systems. During
the late 80's, for instance, a branch line was constructed
from Pecos, Texas, on the Texas and Pacific, up the
Pecos River valley as far as Roswell, New Mexico. This
terminus was only eighty miles by wagon road from
White Oaks. The new railroad was intended to tap the
rich agricultural and cattle-raising resources of the adja-
cent territory. As soon as C. B. Eddy, the builder, had
completed his line, he became aware of the possibilities
of a more direct route from western Kansas to El Paso,
the main gateway into Mexico, and he transferred his
interest to the new project. Work was started at both

— *Lina Parker Mathews*

MR. AND MRS. MORRIS PARKER IN LATER LIFE

ends, and by May, 1890, ten miles of the road had been completed north from El Paso. It was known as the White Oaks and El Paso Railroad.

For years thereafter, progress was slow because of intense opposition on the part of both the Southern Pacific and Santa Fe railroads. The aridity of the territory and the prospect of slight local tonnage discouraged outside capitalists from investing, and the depression of 1893 added another difficulty. Eight years (1890-1898) passed before the road reached the Pecos at Santa Rosa from the north and Alamogordo, eighty miles above El Paso, from the south. White Oaks was left in the middle, waiting to become a railroad town.

In the meantime the coal fields at Capitan, east of White Oaks, and the timber resources on the high plateau of the Sacramento Mountains, ten to fifteen miles east of Alamogordo, were being exploited. Coal and timber brought additional tonnage to the railroad and added an inducement to capitalists who might contribute the money necessary to complete the last link.

In 1902 the final spike was driven, but White Oaks was left off to one side, high and dry. A new town, Carrizozo, was laid out on the wide-open prairie twelve miles west, all because the leading citizens of White Oaks refused to cooperate. The original survey called for White Oaks as an objective. It was the only town between Santa Rosa and El Paso and was equidistant between them. The railroad requested that the townspeople provide a right-of-way, free of charge, along a proposed route which would least affect buildings already in existence. It asked for a flat, vacant acreage near the west edge of town for depot site, shops, and sidings, and a cash bonus of $50,000.

Various town meetings were held, and at first the requests met with apparent approval. Later, opposition developed from an unexpected source. The railroad, as I recall, met this opposition by cancelling the bonus. It was not enough. A select few of the leading citizens refused to compromise and expressed determination to make no concessions and to spend not one dollar. Their slogan was: "Regardless of what we do, the road necessarily *must* pass through White Oaks. It is the shortest, easiest, cheapest route with the lowest grade over the summit. There is no other way."[3]

There was another way. The engineer's statements and the survey map showed that a second route was available north of Lone Mountain and the Jicarillas, where the grade over the summit was even lower than at White Oaks.

Even after construction was started on this alternate route, the dissenters still persisted, but now the slogan had changed. It said, "White Oaks, with its manifold advantages, does not need a railroad."

Their mistake soon became obvious. When the Old Abe declined, there was nothing left to support the town. Thus by their own obstinacy and lack of foresight, those whose lives were most closely associated with the welfare of White Oaks hastened its oblivion.

---

[3] William A. Keleher, *The Fabulous Frontier,* rev. ed. (Albuquerque: The University of New Mexico Press, 1962), pp. 278-91, says nothing about the intransigent attitude of the White Oaks citizens. He thinks White Oaks was bypassed because C. B. Eddy needed a line to the Dawson coal fields when the coal mines east of White Oaks petered out. He states that White Oaks had pledged nine miles of right-of-way, forty acres within the confines of the town, and $50,000 in cash (p. 288, n. 6).

Carrizozo is a fine town, with excellent schools, churches, residences, stores, and business enterprises. If destiny had placed it on the plateau-basin twelve miles east in the mountain-forest amphitheatre which surrounds White Oaks, everybody would call it the best town in New Mexico. Instead of a ghost town, White Oaks would be a thriving, contented, and growing city.

Shortly after its completion, the railroad was taken over by the Phelps Dodge Corporation. The name was changed to El Paso and Northeastern. It was consolidated with the El Paso and Southwestern (then building by Phelps Dodge). The two were later made a part of the Rock Island-Southern Pacific transcontinental line, Chicago to Los Angeles. White Oaks, slowly decaying, watched it happen.

Had the citizens of White Oaks been less unreasonable, had they foreseen the result of their stubbornness, Carrizozo would not exist and White Oaks would be a center of trade and industry, producing coal, timber, cement, lime, and building stone, a mountain resort town with summer cottages easy of access.

What it is — it is. What it was remains a pleasant memory in the minds of those who lived there before the turn of the century. To us oldsters who remember those good, golden days at White Oaks, the dreams of better times ahead for that particular spot remain vivid — perhaps more so as the years go by.

Lest We Forget

MINING IS NOT an easy game and it can be a dangerous one. An old mining man has many memories of failure and defeat and even death. An adage which might be applied here says that the really good, successful miner is the one *who knows when to quit*. The old-time miner used to say, "One more foot, one more shot," and he could cite examples of millions of dollars resulting from the application of this philosophy. They are the exceptions, however, that prove the truth of the major proposition.

Mr. Sigafus was the exemplary figure in White Oaks history. He came into town unheralded, paid his way, took but little interest in local affairs, and ran the North Homestake. When his mill was worn out and mine values had decreased to the point that a profitable operation was doubtful, he simply folded up and quit. He did it without fanfare and with no expressed regrets. He went off to California and applied a portion of his profits to another, and better, gold mine near Yuma.

Frank Lloyd, another excellent miner and manager of the North Homestake, suffered a different fate. He was highly respected and popular with his men. His wife and family of five children were beloved by all the townspeople. About 1896 he left White Oaks and became superintendent of the Oro Nolan mine at El Oro west of Mexico City. He was active in the exploration of the great gold deposits of that region. Within a year of his leaving White Oaks, the shocking news of his death affected the entire community. He and one of his workmen were victims of an unusual and tragic occurrence.

They had entered a closed cage — "elevator" in mine parlance — for the purpose of being lowered to mark the water level in the shaft at the start of pumping operations for dewatering. The elevator descended and the two men were lowered beneath the surface of the water and drowned like caged rats. As the cage neared the point where the water was supposed to be, the engineer lowered it slowly. As a result he did not notice when it made contact with the water. Receiving no signal, he continued lowering. Finally he stopped and waited. Still no signal! Suspicious that something was wrong, he raised the cage to the surface, but too late. The two men were dead.

Investigation disclosed that the bell cord, the only means of signaling, had been pulled up out of the water, coiled, and placed on a shaft timber high above the water level. When the cage reached that point, there was no way to signal the engineer to stop.

Mr. and Mrs. Lloyd were fine physical and moral specimens and their children, all of school age, were healthy and well behaved. His loss was a terrible blow.

This fatality brought on more trouble. When Mr.

Lloyd sent to White Oaks for his family, he included an invitation for a man named Ferguson to join him. The whole family, including a son Don, who was a young engineer, became a valued addition to the home life as well as the work of the Lloyds at El Oro. The father had been head amalgamator and master mechanic for Mr. Lloyd at the North Homestake during most of the time that the property was in operation, and under him the son became an efficient, trustworthy engineer, well versed in mechanics.

The tragic death of Mr. Lloyd, his idol, was almost more than he could readily survive, and for several months it was a question whether the young man would regain his faculties.

The permanent residents of White Oaks, of course, died at home, and for that reason the memories of the village center on an acre of ground west of town alongside the road to Carrizozo — the cemetery. The best part of the once active, prosperous community lies there at rest. Along with many who are forgotten, the burial plot holds the remains of men who exerted no inconsiderable influence on the political, cultural, and business life of the region and of the state. Governor McDonald is there with his wife and daughter. So are Mr. and Mrs. Hewitt, Dr. and Mrs. Paden, and Mrs. Susan McSween Barber, known as the Cattle Queen of New Mexico. These and so many more.

This, however, is a cemetery with a difference. Visit a burial ground out West anywhere, particularly if it is near a ghost, or near-ghost, mining town, and one is in the presence of ruin and neglect. The gravestones are leaning or have fallen down. The desert vegetation has

— John Kelt

DAVID L. JACKSON IN 1956
The ruins of White Oaks behind him

taken over. The place is actually a monument to the disregard and forgetfulness of the living.

It is not so with the cemetery at White Oaks.

In the early 1890's two young colored men arrived in town looking for work: Will Adams and David L. Jackson. About a year later Adams left, but Jackson remained — the only permanent colored resident White Oaks ever had. He was still living there in 1945, a respected and substantial citizen of the community.

For a quarter of a century or more he kept up the cemetery, where so many of his friends were buried. Before national holidays, particularly Decoration Day, he repaired the fence, pulled weeds, cut the high grass, and decorated as many as possible of the graves with branches and flowers gathered from the hills and from neighborhood gardens in tribute to the people he had loved and who had loved him.

To Dave Jackson[1] the old-timers of White Oaks and their descendants owe a debt of gratitude. He was one who did not forget.

------

[1] Dave Jackson arrived in White Oaks on the stage in 1897 (*Lincoln County News,* April 27, 1956). He died in Carrizozo in 1963.

The Mines of White Oaks

WHITE OAKS is said to have the deepest *dry* free-milling
gold mines in the United States, probably in the world.
Not until the Old Abe reached a depth of 1,350 feet
was there any evidence of water seepage in any of the
shafts. As the grade of ore at that depth was noncommercial, no attempt was made to sink deeper.

There are seven main working shafts. The North
Homestake has three with depths of 600, 700, and 800
feet. The South Homestake has two of 500 and 1,066
feet; the Old Abe one of 1,350 feet; the Lady Godiva
one of about 500 feet. All are vertical, two-compartment
shafts except the Lady Godiva's, which is vertical for
100 feet, then inclines seventy-five degrees to follow the
vein. The collar (surface) of the Old Abe is 100 to 200
feet lower in elevation than any of the others.

The 400-foot level of the Old Abe corresponds
roughly with the water level of wells in the town, a mile
and a half to the east. That no seepage reaches the gold
zone is due to a distinctive shift in the rock formation,

which changes from sedimentary to igneous with no break, fissure, or crack in the latter whereby water may enter the centralized mine area.

White Oaks gold is "free" — that is, not associated with other metals. The average purity is close to 900/1000 (90%). The remaining ten percent is a consistent silver alloy. At the United States unit value of that period, $20.67 per ounce troy, the bullion or mint value was approximately $18.50 per ounce. In 1935 the value of $20.67 was changed by executive decree to $35.00, the present-day figure. At this valuation White Oaks gold is worth about $31.50 per ounce.

Extracting by stamp mill, crushing to forty mesh, and plate amalgamation recovered 80 percent to 85 percent of the gold content of the ores. Loss was due primarily to coarse screening, to the fact that the sand (ore) was not sufficiently pulverized to liberate the finer particles of metallic gold, and to the failure of small, light particles of gold carried by the mill streams to come in contact with the mercury. Sometimes these small particles were coated (covered) by colloidal elements in the pulp.

Discharged tailings were impounded in settling ponds below the mill. The water thus recovered was pumped back into the mill tank for repeated use. Prior to 1892 it was customary, at intervals, to sluice the impounded tailings down the arroyo to make room for future tailings. Tonnage and values thus washed away amounted to thousands of dollars. Although daily assays recorded these losses, at that time there was no practical method known for recovery of gold from these tailings. After 1892, cyanidation offered a means of extraction and profit.

Total production (my estimate) from the White Oaks mines was approximately four and a half million dollars. The Old Abe was first; North Homestake second; South Homestake third.

The geology of the region (I write from memory) is interesting. The summit of Baxter Mountain and the escarpment facing east is composed of Cretaceous sediments with many well-defined fossils. The uplift of the mountain range and the break of the sediments were caused by volcanic action, a surgence of siliceous-feldspathic rock: *felsite-porphyry*. Axis of the uplift: north-south.

Within this widely exposed porphyry, near the base of the escarpment and parallel to it, is an intrusive dike, perhaps better classified as an elongated volcanic plug, of diorite, locally referred to as augite-syenite, about three quarters of a mile in length and varying in width with an average of about 300 feet. The south end, near the end line of the South Homestake, terminates abruptly, bulges eastward, and makes a well-defined fusion contact with the porphyry. To the north it disappears under hill-slope debris of mixed porphyry and metamorphosed sedimentaries. The fissures within this mineralized zone are due, no doubt, to the cooling and settling of this diorite dike, or plug. Hot solutions and vapors from the underlying magma are responsible for the veins, vein fillings, and gold ore bodies.

On each side of the diorite, and about 800 feet apart, are two remarkably perpendicular ore-bearing fissure veins. The west vein is entirely in felsite porphyry and contains two separate ore bodies or "shoots" covered by the North and South Homestakes. The east vein, covered by the Old Abe, contains one ore shoot and an auxiliary wall or off-side "pocket" known as the Fish Pond. The enclosing rock formation is more complex than that of the west vein or of other deposits in the White Oaks district. It consists of porphyry similar to that of the west vein plus a disrupted portion of contact-metamorphosed sediments alongside.

Another ore body, known as the Devil's Kitchen, lies entirely in the diorite and is covered by the South Homestake claim.

Thus there are five separate and distinct known gold-ore deposits in the three principal producing properties. A detailed description of each mine follows in the order of its discovery and the start of production.

SOUTH HOMESTAKE — "The Devil's Kitchen" ore body is half way (250 feet) up the south hill slope of a west-east gulch with "stock work" near the surface and a north-south fissure below — all in diorite. The glory hole is an open pit excavated to a depth of seventy feet. Beneath it a 600-foot shaft was sunk, later destroyed by fire. Workings were reopened by a new shaft from the surface above the Kitchen to a depth of 700 feet.

The perpendicular walls of the open pit show the same homogeneous diorite on all four sides with no evidence of a vein. Obviously the shattered outcrop of gold-bearing quartz was confined to this area, about 100 feet square.

Below the Kitchen the seams unite, forming lenticular veins 100 feet and more in length, three to five feet in width, remarkable in their direct vertical extension. The vein filling consists of inclusions of wall fragments, coarse crystalline quartz, calcite, and other accessory earth minerals. Associated with them are frequent good-sized deposits of huebnerite,* a manganese tungstate characteristic of the basic diorite and not found in the other (felsite-porphyry) ore bodies.

The ore lenses increase in size to a depth of about 500 feet, after which there is a gradual diminution. The lowest level is not profitable on account of excess wall rock which has to be broken to extract the gold-bearing quartz.

A third shaft 800 feet deep was sunk on the opposite hill slope near the ridge about 500 feet (slope measure) from the gulch and 300 feet from the dividing line of the North Homestake. This work is all on the fissure vein in porphyry west of the diorite. The vein, four feet and more in width, extends north, the full length of, and beyond, the North Homestake ground. The ore shoot, outcropping on the North Homestake, has a decided rake to the south, entering the South Homestake at a depth of 275 feet. On account of the wide spreading of the vein in depth, and the decreased gold value per ton of rock mined, the ore shoot below the 800-foot level was too low grade for profit.

NORTH HOMESTAKE — The workings are all in porphyry and on the same vein mentioned in the preceding paragraph.

The main shaft, vertical, is 1,066 feet deep. The ore shoot, an elliptical "pipe" or "chimney," is near the north end line of the property. The average cross section is fifteen feet wide and seventy feet long. Smaller near

---

*Because of its jet-black color, huebnerite is often confused with and referred to as wolframite, a closely related tungstate with a varying percentage of iron (Morris Parker's note).

the surface, it increases both in width and in length with greater depth. Commercial values extend into the wall rock on both sides of the vein proper.

There is a second shaft, 500 feet deep, on the same vein near the south end line. The ore shoot has a fairly uniform width of four feet. Length near the surface is about fifty feet, increasing to 300 feet at depth. The shoot rakes south, entering the South Homestake at a depth of 250 feet. At a depth of 500 feet, it is entirely inside the South Homestake.

The 800-foot South Homestake shaft was sunk to intercept and mine this ore shoot. At the 400-foot level the width was four feet and average value was ten to twelve dollars per ton. On the bottom level (South Homestake) at a depth of 800 feet, it was ten feet more in width and the values less than three dollars.

OLD ABE — This was the last in operation — the biggest and best of them all. There was only one working shaft, 1,350 feet deep — in porphyry east of the diorite. Situated almost due east of the main shaft of the North Homestake, it was about 800 feet distant from it.

The Old Abe was a true fissure vein. The ore shoot, narrow and short near the surface, widens to a fairly uniform width — four feet to a depth of 500 feet. From this point there is a gradual widening to ten feet or more at the bottom level. In like manner the length increases to about 400 feet.

The levels (drifts) south from the shaft — never extended more than a limited distance — would, if continued straight ahead on the known course of the vein, pass into the south-end diorite "bulge" mentioned above.

To the north, certain of the drifts passed from the porphyry into the contact zone of metamorphosed sediments in that direction. Surface exposure of the hill slope east and north of the shaft is composed entirely of these stratified clays, shales, and slates — fair enough evidence that the felsite-porphyry in which the main ore body occurs is a segment or resection of the volcanic intrusion that caused the mountain uplift.

Short crosscuts west toward the diorite at depths of from 500 to 700 feet disclosed an irregular elliptical ore body, the "Fish Pond," from which several thousand tons of good ore were mined. It is said to be the most productive area, for its size, in the district. The rock formation in which it occurs is sedimentary, highly metamorphosed and brecciated. It is the only ore body of any consequence in any of the White Oaks mines known to occur in this formation.

The presence of sedimentaries at this point can be accounted for in two ways. They either form an "island" suspended in or enclosed by the intrusives (diorite and porphyry) or a "finger" — peninsula — pendant from the main mass of sediments no great distance to the north.

Brecciation was due to the squeeze pressure of the intrusives. Metamorphism was the result of both pressure and heat, the latter so intense that the sediments were changed to so-called hornfels — a fine-grained compact mass in which no phenocrysts (large crystals) are discernible by the naked eye. This particular area provided an ideal receptacle for the subsequent infiltration and deposition of minerals, in this case the most part with lime spar and gypsum. The coating and filling of the interstices of the dark hornfels with white crystals of lime minerals is a distinctive feature of the Fish Pond deposits. A number of beautiful and rare specimens of native gold were found imbedded inside the transparent crystals of both calcite and selenite.

LADY GODIVA — This is a fractional claim between the North Homestake and the Old Abe. It is a seventy-five-degree-inclined fissure vein in diorite. The apex is east of the center of the dike, the dip west. The shaft is 500 feet deep and is directly west of the Old Abe. The vein is narrow but strong with infrequent bunches of commercial ore. There was a small amount of drift work; no crosscutting.

There is no apparent vein or fissure connection between the Lady Godiva and the Devil's Kitchen ore bodies, though both are in the diorite.

LITTLE MACK — About a mile north and in line with the Homestake west vein, there is another separate and well-defined true fissure vein, four feet in average width, in granite-porphyry. Some commercial ore has been mined. It is of interest that the vein filling shows an appreciable presence of manganese in oxide form and also associated with iron. Later work proved disappointing.

The two Homestakes, the Old Abe, and the Lady Godiva — all in one group — cover an area less than one-half mile long (2,500 feet), less than half that in width, and less than one-eighth of a square mile.

All three veins are the result of injections from below — a forced opening (widening) of fracture planes by the intrusion of hot solutions and vapors (gases) from an underground magma containing disseminated (low grade) gold and other minerals.

The known length of each of the two veins in felsite-porphyry is approximately 2,000 feet. Both are strong, well-defined fissure veins with a fairly constant width of four feet. There is a gradual spreading at increasing depth alongside the channels of ore deposition (ore shoots). The entire vein filling, in addition to the ore shoots, contains some gold worth fifty cents plus per ton — appreciably more than the enclosing walls. To what depth this extends, of course, is not known.

Form, position, and character of the ore bodies (shoots) strongly indicate that they are the result of individual geyser-fumarole action, meaning that they originated in "a small hole from which volcanic vapors issue." The confined limits of the commercial ore, the inverted funnel shape and perpendicular channels of deposition and exit are plain evidence justifying this conclusion.

Geyser activity — its violence and effect — is perhaps best shown in the ore shoot of the North Homestake's main 1,066-foot shaft. Width of ore near the surface, three feet, value per ton, $30.00; width at lowest level, thirty-five feet, value per ton, $4.00 and less.

Below the 200-foot level there is no evidence of the well-defined demarcation line (slickensides and gouge) which so plainly separates the vein and enclosing porphyry (wall rock) in all other places. In other words, the former walls of the vein have been obliterated. Geyser action is the only explanation: confinement in length and expansion in width; forceful penetration and deposition of gold in the enclosing wall rock rather than in the longitudinal extension of the vein at either end of the ore channel.

The ore was mined to the outer limits of commercial value. Beyond this, deposition continues in amounts not profitable to mine.

This same spreading of the vein with depth, increase of volume (tonnage), and corresponding decrease in gold content per cubic foot or ton is apparent in both of the felsite-porphyry veins. That the Devil's Kitchen deposit, in diorite, shows these conditions in lesser degree is due to the fact that there is no vein beyond the limits of the ore channel. In the Lady Godiva no defined ore shoot of commercial value was uncovered.

The various localized ore shoots unquestionably are the result of emanations, relief vents, from the same underground reservoir of magma, and were subsequently deposited by cooling and relief of pressure. This caused the enrichment of the area affected by the geyser — hence the ore shoots.

The same is true of the gold-huebnerite in the Devil's Kitchen ore body: the gold originated in the same underground magma; the huebnerite derived from segregation in a confined area of the more basic diorite. The channel of egress was not connected with the channels of the acidic felsite deposits.

All the gold of the district is of the same fineness, approximately 900/1000, evidence of the same genetic origin. There is, however, a notable difference in the specimen ore. All veins produced both granular (coarse) and fine, disseminated particles, but those of the felsite showed a much higher percentage of crystalline "leaf" and "wire" gold, whereas in the diorite the gold, as a rule, was in coarse, heavy grains tightly enclosed in narrow seams of hard, crystalline quartz — frequently fifty percent or more pure gold, capable of being cut and polished for pendant jewelry, watch fobs, stick pins, and so forth. Specimens of bright, sparkling gold were sometimes also found in the huebnerite lenses, especially on the outer rims.

HUEBNERITE — This mineral, found only in the diorite, has a specific gravity of 7.2, three times heavier than quartz, and was a prime nuisance in mill amalgamation —dragging, hanging back, and covering the copper plates, necessitating frequent flushing with a hand hose. Many tons were crushed and washed down the creek with South Homestake tailings. In the mine workings it often occurred in solid masses filling the entire vein fissure, two to four feet in width. Again, it was found in alternating bands with quartz. Because of the difficulty in milling, such areas as a rule were left standing in the drifts and stopes.

We knew nothing of its use or value, to our sorrow later on. The composition is tungstic oxide, 75.6 percent; iron oxide, .40 percent; manganese oxide, 24 percent. The main use for tungsten is for hardening steel, for high-speed self-tempering tools, and for wires for incandescent lamp filaments. It is also used for contact points in spark plugs and coils and for telegraph relays and voltage regulators. When World War I started in 1914, the market value of tungsten ore was $3.00 to $5.00 per unit (percent) based on sixty percent tungstic oxide. Most of the U. S. production came from Boulder County, Colorado. During the war the price jumped to $80.00 per unit, which, with the bonus for content over sixty percent, gave to the higher-grade ores a market value of as much as $5,000 per ton — sometimes more. Known deposits theretofore ignored entered production along with new ones, and the profits were immense. The common form of contract was $40.00 per unit for ore or concentrates showing fifty percent tungstic oxide, plus twenty cents per unit when the showing was over fifty percent. The bonus was sometimes raised by competitive buyers. White Oaks huebnerite, showing seventy percent plus, thus had an approximate value of $2,900 per ton.[1]

In 1918 a couple of war-metal scouts visited White Oaks, leased the South Homestake, presumably for gold, on a small royalty basis, mined the exposed huebnerite ore and, by using the assay sample grinding machinery and a hand-made jig, cleaned up a fortune. This heretofore worthless, detrimental mineral was thus given away to outsiders and sold by them at a minimum of $2,500 per ton, verifying the old adage: It Pays to Investigate.

A feature of the White Oaks ore bodies, common to many Tertiary gold deposits of the West, is the absence of other minerals such as silver, copper, lead, and zinc. Pyrite (iron sulphide) is so rare as to be negligible. In hand panning (mechanical concentration of crushed vein material in or outside the ore zones), however, there is noted a small amount of "black sand" — iron oxide (magnetite) — which is the only heavy mineral other than gold and the huebnerite mentioned above. It was reported that in a winze sunk below the 1,350-foot level of the Old Abe, a noticeable amount of arsenical iron sulphide was encountered. Whether the report was true or false, I do not know.

It would appear that the theory of "liberation" of gold by surface oxidation has but little confirmation in the deposits at White Oaks. The gold, as mined from

---

[1] The El Paso *Times* for July 11, 1969, carried a story describing a new project, a "hydro gravity ore mill," just going into operation south of Carrizozo. "Ore to be milled for the first year will come from White Oaks mine dumps, starting with South Homestake mine. . . . The most valuable product of the operation will be tungsten."

the ore channels (shoots), was apparently deposited where found without any subsequent change or release from association with any other mineral. The absence of ferrous iron as coloring or rust is a strong argument. No minerals were present to oxidize. The tailing ponds have the color of common sand, earthy.

Is there any gold left at White Oaks? I think there is. Mr. Frank Keller, after making quite an exhaustive investigation, as noted earlier, *reported favorably*, even though the tonnage and values did not cover the price asked. Why the favorable report except for some convincing evidence of values not recognized by the owners? The deal fell through not because of Keller's report but because outside promoters, not in on the deal, operated insidiously to defeat it.

The fact is that little or no intelligent prospecting, following geologic clues, has ever been done. During the summer of 1892 R. C. Hills, chief geologist for the Colorado Fuel and Iron Company and one of the most capable geologists of his time in the United States, made a reconnaissance survey of the region between Santa Rosa, New Mexico, and El Paso to determine the mineral resources on both sides of the valley and estimate prospective tonnage for the railroad. The party consisted of seven or eight men. I had just returned from the School of Mines at Rolla, Missouri, and was employed as assayer for the group with a portable outfit to make assays as we went along. My brother Frank was also one of the party. The journey lasted a month or more, and we examined and sampled every mine and prospect hole along the proposed route.

Mr. Hills was a kindly man, talked freely and understandingly in terms we could follow, and classified rocks and fossils in a manner which remained an inspiration to me as I engaged in similar work in later years. He could have done much to unravel the geology of White Oaks, but since the deposits of coal and iron alone offered tonnage for the railroad, he made no examination of the gold mines.

It is deeply regrettable, I often think, that no high-class geologist, so far as I know, has examined the rock structure and the ore deposits of the White Oaks gold area with special attention to their origin. During the exploitation of the known deposits, nobody cared, and now that the workings are inaccessible, the diagnosis would be far more difficult. The difficulties would not be too great, however, for the drawing of intelligent conclusions. The relative deposition and association of gold in the felsite and diorite formations surely have some distinctive meanings, and are worth study.

Some day the study may be made and White Oaks will rise again.

Terms Used in Mining[1]

AMALGAMATION — A process for the extraction of
gold and/or silver from their crushed or pulverized
ores, in which mercury is added to the ore pulp forming an alloy (amalgam) of the precious metals and
the mercury. The amalgam is then removed from the
pulp by any one of several methods, after which
the mercury is frequently expelled by retorting.

ARRASTRA (from Spanish *arrastrar*, to drag) — An
apparatus for grinding and mixing ores by means of
a heavy stone dragged around upon a circular bed;
it is used chiefly for ores containing gold.

BRECCIA — A fragmental rock whose components are
angular (not waterworn as with conglomerates) and
embedded in a fine-grained matrix. Brecciation is the
process by which such rock is formed.

---

[1] For more detailed explanations, see Albert A. Fay, *A Glossary of the Mining and Mineral Industry* (Washington: Government Printing Office, 1920; reprinted 1947).

BULLION — Refined silver or gold, usually in the form of bars or ingots.

COMPRESSOR — A machine for compressing air in order to operate machinery or ventilate a mine.

CONCENTRATING TABLE — A table on which a stream of finely crushed ore and water flows downward and the heavier metallic minerals lag behind and flow off into a separate compartment.

CRADLE AND ROCKER — A trough in which gold-bearing sands are combined with water and swung back and forth, bringing the particles of gold together at the bottom.

CUPEL — A small, shallow, porous cup used in assaying gold and silver ores.

DRIFT — A horizontal passage or tunnel which follows a vein, as distinguished from a cross-cut which intersects the vein.

FIRE ASSAY — The assaying of metallic ores, usually gold or silver, by methods requiring a furnace heat — and generally including such processes as cupellation — whereby any gold or silver is actually separated from the ore.

FREE GOLD — Gold as discrete particles uncombined with other minerals.

FREE-MILLING ORES — Ores containing free gold or silver which can be extracted by crushing and amalgamation, not requiring treatment with chemicals.

GLORY HOLE — A large open pit from which ore is extracted by dropping it through shafts or winzes connecting with a haulageway driven beneath the ore body.

GOUGE — Finely abraded material occurring between the walls of a fault, the result of grinding movement; a layer of soft material along the wall of a vein, favoring the miner by enabling him, after "gouging" it out with a pick, to attack the solid vein from the side.

GYRATORY CRUSHER — A rock crusher consisting of a vertical spindle the foot of which is mounted in an eccentric bearing within a conical shell. The top carries a conical crushing head revolving eccentrically in a conical maw.

HOIST — The engine used to operate the cage or elevator in a mine.

JIG — An apparatus in which ore is concentrated on a screen or sieve in water by a reciprocating motion of the screen, or by the pulsion of water through the screen.

LEACHING — The separation of metals from ore by allowing a solvent such as acid to percolate through the material.

LENTICULAR — Lens shaped; a mass thick in the center and thinning at the edges.

MAGMA — Naturally occurring molten rock material generated within the earth, from which metal-bearing solutions frequently emanate and form ore deposits.

ORE SHOOT — A generally higher-grade streak or portion of a vein of mineral-bearing rock.

RETORTING — A process for removing mercury from an amalgam by heating it in an iron retort until it becomes volatile and can be conducted away.

SHEAVE — A grooved pulley.

SLICKENSIDES — A smooth surface at the side of a vein or fault produced by the movement of the walls against each other.

SPECIMEN ORE — Spectacular or valuable pieces of ore.

STAMP MILL — An apparatus in which rock is crushed by descending pestles, or stamps, operated by water or steam power. A ten-stamp mill thus has ten stamps or pestles.

STOCKWORK — An ore deposit of such form that it is worked in floors or stories — usually a rock mass so interpenetrated by small veins of ore that the whole must be mined together.

STOPE — An excavation from which the ore has been extracted either above or below a level in a series of steps.

TELLURIDE ORE — Ore, usually of gold or silver, in which the precious metals are combined with tellurium.

WINZE — An opening connecting a level in a mine with another level below.

Bibliography

Carson, Xanthus. "When Lincoln County Ghost Town Roared with Life," *Lincoln County News,* May 29, 1969; reprinted from *True West.*

Fay, Albert A. *A Glossary of the Mining and Mineral Industry.* Washington: Government Printing Office, 1920; reprinted 1947.

Fulton, Maurice Garland. *History of the Lincoln County War.* Robert N. Mullin, ed. Tucson: University of Arizona Press, 1968.

Haynes, Lydia. "Mill to Produce Tungsten from old Gold Mine Dumps," El Paso *Times,* July 11, 1969.

Hough, Emerson, *Heart's Desire.* New York: Macmillan, 1905 (copyright, Out West Publishing Company, Curtis Publishing Company, 1903).

Jones, Fayette A. *New Mexico Mines and Minerals.* Santa Fe: New Mexican Printing Company, 1904.

Keleher, William A. *The Fabulous Frontier: Twelve New Mexico Items.* Rev. ed. Albuquerque: University of New Mexico Press, 1962.

Murbarger, Nell. "Ghost of Baxter Mountain," *Desert Magazine* 15 (November, 1952): 4-8.

"Ramblin' around Lincoln County. Tape-recorded interviews with People Who Helped Build Lincoln County," *Lincoln County News,* April 27, May 4, 11, 1956 (interviews with David L. Jackson).

Rasch, Philip J. "Who Killed Who in Lincoln County, N.M.," *Lincoln County News,* July 17, 1969.

H.S. "White Oaks: The Greatest Mineral Region of New Mexico," El Paso *Times,* April 23, 1887.

Sinclair, John L. "Little Town of Heart's Desire," *New Mexico* 18 (December, 1940): 18-19, 39.

Stanley, F. *The White Oaks Story.* Published by the author, n.p., n.d.

Stivers, C. E. *White Oaks, New Mexico.* El Paso: Press of the Daily News, 1900 (distributed by Paul Mayer, Livery, Hay and Grain, White Oaks).

Wetzel, Charles L. "Mining in White Oaks, New Mexico." Course paper, English 3102, Texas Western College, May 20, 1957.

White, Marjorie. "Aura of Golden Years Surrounds White Oaks," El Paso *Times,* August 24, 1969.

Newspapers Consulted

El Paso *Times*

El Paso *Bullion*

Lincoln County News (Carrizozo)

Rocky Mountain News (Denver)

Socorro *Bullion*

White Oaks *Eagle*

Index

147